The Author

David Dickson was born in 1947. He held
scholarships at Westminster School and Trinity
College, Cambridge, where he read mathematics,
and after graduating spent two years as a medical
journalist. He subsequently became general
secretary of the British Society for Social
Responsibility in Science. His main concern has
been with the mutual relationships of science,
technology, art and society, and he has recently
compiled a study for Unesco on the national and
international organization of scientific research.
At present he is science correspondent of the
Times Higher Education Supplement.

The Series Editor

The series is edited by Jonathan Benthall, author
of *Science and Technology in Art Today*, editor of
Ecology, the Shaping Enquiry, *The Limits of Human
Nature*, and co-editor of *The Body as a Medium of
Expression*. Formerly lecture programme organizer
at the Institute of Contemporary Arts, he is now
Director of the Royal Anthropological Institute.

Technosphere

Technosphere is a new series that presents individual
studies of particular sciences and technologies in
terms of their human repercussions. The series
will include original analyses of a wide variety of
subjects – transport technology, drugs, computer
technology, operational research and systems
analysis, photography, urban planning, film,
alternative medicine, and so on – with the
emphasis in each case on the present state of the
science or technology in question, its social
significance, and the future direction of its
probable and necessary development.

Also available in this series

Television Technology and Cultural Form
Raymond Williams

Alternative Technology

and the Politics of Technical Change

DAVID DICKSON

FONTANA/COLLINS

First published in Fontana 1974
Second Impression March 1976
Third Impression May 1977

Copyright © David Dickson 1974

Made and printed in Great Britain by
William Collins Sons & Co Ltd Glasgow

Contents

Acknowledgements

Writing a book of this nature has incurred many debts. In particular, I would like to acknowledge the advice and comments received, among others, from Robin Clarke, Dave Elliott, Chris Green, Dorothy Griffiths, Mike Hales, Chris Harlow, Peter Harper, Shivaji Lal, Thierry Lemaresquier, Jonathan Rosenhead, David Wield and Robert Young. Also Jonathan Benthall, who originally suggested a book on this subject, Pru Richardson-White – now Pru Dickson – who helped patiently with much of the typing, and members of the British Society for Social Responsibility in Science, Project 84 Arts/Science Group, and the editorial collective of *Radical Science Journal*, many of whose ideas have strongly influenced my own, in both conscious and unconscious ways. Errors and misinterpretations, of course, remain my responsibility alone.

The revolution which is beginning will call in question not only capitalist society but industrial society. The consumer's society must perish of a violent death. The society of alienation must disappear from history. We are inventing a new and original world. Imagination is seizing power.

Sorbonne, Paris, May 1968

Introduction

Contemporary society is characterized by a growing distrust of technology. The many social benefits which technology has helped to bring about are being increasingly counterbalanced by the social problems associated with its use. These range from the oppression and manipulation of the individual to the widespread destruction of the natural environment and the depletion of the world's finite supply of natural resources. At the same time, man's technological skills have so far failed to provide effective solutions to many of the world's major problems, in particular those of mass poverty, starvation and international conflict. Technology is no longer seen as the omnipotent God that it was even ten years ago, when the British Labour Party won a General Election on the pledge to create a 'white heat of technological revolution'.

Yet we cannot escape the fact that technology has become an integral part of our social world and an essential element in almost every field of everyday activity. One of the particular characteristics of modern life is that some form of machine is required to carry out almost any act within the social domain. We use machines to travel, to communicate, to produce commodities, to provide services and even to entertain each other. An understanding of the role of technology in society has become important, not just for its own sake, but as part of our understanding of society itself.

The purpose of this book is to discuss both the social function of technology, and the particular socially-accepted interpretations or 'legitimations' which are placed on this function. I have attempted to show how the contemporary problems associated with technology stem as much from the nature of technology as from the uses to which it is put, but that the former is largely determined by social and political factors, of which technology can never therefore be considered independent. I hope to indicate how an alternative technology might be developed which would avoid the problems associated with contemporary technology that we now experience. But I also emphasize the political obstacles that stand in the way of any attempt to put an alternative technology into effective practice.

My general thesis is that technology plays a political role in society, a role intimately related to the distribution of power and the exercise of social control. It does this, I maintain, in both a material and an ideological fashion, implying that in both senses technological development is essentially a political process. At a material level, technology sustains and promotes the interests of the dominant social group of the society within which it is developed. At the same time, it acts in a symbolic manner to support and propagate the legitimating ideology of this society – the interpretation that is placed on the world and on the individual's position in it. Both material and ideological factors, I suggest, play an important part in determining the nature of technology itself.

Thus while arguing against those technological determinists who see social developments – including changes in social organization and in the distribution of power – spontaneously emerging from developments in technology, I hope to indicate that it is equally possible to challenge those who support a purely economic determinism, in which technology is seen as no more than a neutral tool of economic, or even political, development. I would argue that the social relations of production – the relationships between the various social groups or classes involved in the production process – become reflected in the means of production, in other words that technology and social patterns reinforce each other in both a material and an ideological fashion. 'Most new industrial technologies are found because they are sought' was one of the main conclusions of the historian of technology, Jacob Schmookler, after many years of study of the process of technological innovation.[1]

The implication of this thesis is that one can only understand the nature of the technology developed in any society by relating it to the patterns of production, consumption and general social activity that maintain the interests of the politically dominant section of that society. Within capitalist societies, such interests are maintained by hierarchical patterns of social organization and accompanying authoritarian forms of social control. The validity of collective experience is denied; political power is consolidated in the hands of the few at the top, and fragmented between the many at the bottom. The dominant modes of hierarchical organiz-ation and authoritarian control, I maintain, become incorporated in, and hence come to coincide with, the technology that is de-veloped by capitalist societies. So, too, is a predominantly func-tional attitude towards the natural environment that amply fulfils the prediction of Descartes that science would provide the knowledge 'by which we may be able to make ourselves masters and possessors of nature'.[2]

I would also suggest that many nominally-socialist countries, by appropriating and subsequently developing a mode of pro-duction initially formulated within a capitalist framework, have been obliged to introduce forms of social organization and control that are essentially capitalist in nature in order to make effective use of this technology. Thus although my critique relates pri-marily to the situation in the industrialized capitalist countries – as well as to the nascent capitalist economies of many under-developed countries – much of it could equally be applied to industrialized nations of a nominally-socialist nature.

It is from this general perspective that I wish to discuss the problems associated with contemporary technology, and in par-ticular the various proposals that have been put forward as to how such problems might be resolved through the design of an alternative technology. This technology would embrace the tools, machines and techniques necessary to reflect and maintain non-oppressive and non-manipulative modes of social production, and a non-exploitative relationship to the natural environment.

In the period since the Industrial Revolution, and in particular over the past fifty years, it seems to have become generally accepted that increasingly advanced technology offers the only realistic possibilities of human progress. Technological develop-ment, initially concerned with the improvement of traditional craft techniques, and subsequently embracing the application of abstract knowledge to practical problems, has promised to lead society along the path to a bright and prosperous future. 'There is one powerful force working constantly towards a greater output per head and ever-rising standards of living. It is the march of science and invention; in technical progress lies the economic hope of mankind,' wrote the economist Frederic Benham.[3] The development of technology has even served as an indicator of the general progress of social development; societies still tend to be judged advanced or backward according to their level of technological sophistication.

Such uncritical faith in technology is now beginning to falter. Fuel shortages and power cuts have made man aware of the precariousness of his technological existence. Weapons of mass destruction have provided a sinister back-cloth against which international power struggles are acted out. The individual in contemporary society feels himself increasingly trapped by powerful forces outside his control. He is reduced to little more than an economic cipher, continuously and uncomprehendingly manipulated within a vast, inhuman complex. Technology, originally developed as a means of raising man above a life of poverty, drudgery and ill health, now shows its other face as a major threat to his sanity and his survival. Not surprisingly, many have begun to feel that our technological society has opened the real Pandora's box, and is finding itself rapidly overcome by the contents.

Given this situation, the search for an alternative technology which would avoid such problems appears an eminently sensible one. The use of solar energy for heating houses or methane gas for powering motor cars promise sources of power that do not rely on the continuously diminishing supply of the world's fossil fuels – or the potentially hazardous implications of nuclear energy. New forms of intensive agriculture, based on organic, rather than artificial, fertilizers, for example, and often using techniques discarded many years ago in the interests of 'progress', provide an alternative to a heavy reliance on imported or artificially-stimulated food, as well as preserving the biological integrity of the soil. Small-scale craft industries promise a return to a community-based mode of production, in which every man or woman is in direct control of his or her own life, and work becomes reintegrated with all other areas of collective activity and experience.

Such utopian visions can only increase in attractiveness as the problems associated with our existing technology continue to grow.

While an alternative technology based on these principles may well be a necessary prerequisite for creating a non-alienating, non-exploitative way of life, the development of such a technology is not, I maintain, sufficient to ensure that this state of affairs will be brought about. Indeed, once the political nature of contemporary technology has been realized, it will be seen that a genuine alternative technology can only be developed – at least on any significant scale – within the framework of an alternative society. The achievement of this is a political task. The struggle for emancipation from an apparently oppressive and manipulative technology coincides with the struggle for emancipation from oppressive political forces which accompany it. To argue that technological change is *per se* able to bring about a more desirable form of society is technological determinism carried to utopian extremes.

This book falls into four parts. The first chapter is an attempt to describe as objectively as possible some of the social problems associated with contemporary technology in both industrialized and underdeveloped countries. The second two chapters deal with the way that we look at technology and its role in society. The first of these describes various interpretations of the process of technological development, and argues that any interpretation which sees this process as being independent of political considerations is an ideological distortion of the history of technology; the second attempts to indicate how technological innovation has acted, and continues to act, in a directly political fashion.

The following three chapters discuss two particular types of alternative technology. The technical aspects of a possible utopian technology, stressing in particular how such a technology might be designed along ecologically-stable lines, are described in some detail. An analysis of the theory and practice of intermediate technology – an alternative technology for the Third World – highlights the impossibility of separating technological from political change. In the last chapter, I have attempted to draw together the more theoretical arguments of the book into a composite description of the political function of technology in contemporary society, the legitimating role of science, and some of the implications of both.

Finally, a word of caution. This book is written as an *interpretation* of the role of technology in society. It is not meant as an *explanation* of this role. My thesis that technological change is a political process immediately implies that any explanation requires a political analysis of the contemporary State, and this is not a task which I have set out to do. My intention has been to describe the appearance of technology from a certain perspective, in the hope that others will feel inclined to share my point of view. It has not been to provide arguments that are meant to be taken as a 'proof' of any case (itself a quasi-scientific concept that can be challenged). Nor have I attempted to avoid the charge of eclecticism by adopting a rigid and definitive stance, as it appears that there are many interpretative questions yet to be resolved. This book is offered as a contribution to the discussion of the nature of contemporary society, and will, I hope, help to clarify why we look at and experience technology in a particular way. It is meant to be read against a background of political analysis which can only be explored by turning to various books mentioned in passing references. And as the message of such books makes clear, mere interpretation is not enough. The real task ahead is to change our situation, and this, as I have tried to show, is a political, as much as a technological one.

1 The Case against Contemporary Technology

Reactions against technology are not new. Throughout history man has been warned that he was creating forces he would be unable to control, that machines would eventually take over the planet and demand the total obedience of the human race (if, indeed, it was still allowed to exist), that to place one's faith in science and technology was to make a pact, like Faust, with the devil. St Augustine warned that 'for the injury of man, how many kinds of poison, how many weapons and machines of destruction have been invented'. The Industrial Revolution faced many critics who despaired at the growing importance of the new machines, and at the social problems they seemed to bring with them. Opposition to reason and to rationality, frequently embracing attacks on science and technology, has in particular been experienced by societies that have suffered a major upheaval or catastrophe. The defeat of France in the Napoleonic Wars is claimed to have been followed by a period of rampant mysticism, including a wide resurgence of interest in astrology. Oswald Spengler captured the imagination of a defeated Germany in 1920 with the publication of his *Decline of the West* and his prediction that 'Faustian man will be dragged to death by his own machine'. Aldous Huxley sketched out the blueprint of his nightmarish *Brave New World* during the depression of the early 1930s. George Orwell's *Nineteen Eighty-Four*, an equally damning indictment of the totalitarian possibilities contained in advanced technology, was written in the years immediately following the Second World War.

Whatever one feels about these earlier critiques, the current attack on technology is too serious and too profound to be dismissed as a transient phenomenon. Our society depends as no

other society before it on the efficient running of a vast technologically-based machine. Furthermore, the present mood of dis-
enchantment has emerged as much from direct experience of the
growing social problems associated with contemporary technology
as from consideration of the relationship of technology to abstract
concepts such as humanity or progress. One can attack the
increasing domination of 'scientific' over aesthetic or humanistic
values, or quasi-religious deference to technical expertise. But the
poisoning of one's local river by industrial effluent, the lung-
disease induced by working with asbestos over many years, the
sclerosis of our cities caused by the private motor-car, or the sight
of a Vietnamese child permanently scarred by the pellets of a
fragmentation bomb, bring the same points home with much
more force.

Many conventional discussions of such problems assume what
is often referred to as a 'use-abuse' model of the social function of
technology. They imply, in other words, that the social problems
associated with technology stem from the uses to which technology
has been put, and not from the technology itself, which remains
blameless. Problems like those above are seen either as the intended and often consciously-harmful effects of the attempts of
one individual or group to impose its will on another, or as the
purely accidental side-effects of economic or political processes.
It is this interpretation of the social problems associated with
technology that I wish to challenge. I hope to indicate the reasons
why this model is inadequate to interpret the essential nature of
the role of technology in society. In its place I suggest how we
must see technology itself as part of the political process – even
though isolated machines may play a neutral role in this process –
and hence the problems associated with it as resulting as much
from the nature of technology as from the way in which it is
used.

First, some definitions. In the way that the term 'society' is con-
ventionally used to indicate not only a collection of individuals,
but also the relations between them, so 'technology' will be defined
as an abstract concept embracing both the tools and machines
used by a society, and the relations between them implied by their
use. No distinction need therefore be made for the purposes of this
book between machines and tools, both of which can be defined

as objects selected or fabricated by man as a means of changing the state of his material environment. The above definition allows us to see technology, in common with the legal or the education system, as a social institution. It can also be distinguished from 'technique', which will be taken as the act of applying knowledge, whether directly or with the aid of a tool or machine – that is with the aid of an element of technology – to a particular task.[1]

If we look in detail at the types of problems associated with contemporary technology, we should perhaps start with the problems of environmental pollution, and with the depletion of many of the world's non-renewable natural resources. These are by no means the only, nor, some might argue, the most important problems. They are nevertheless ones that have received a considerable amount of attention over the past few years and there is little need here to underline their severity. Inasmuch as environmental problems are often shared by countries that have adopted the same patterns of industrial development, but under very different political banners, it is possible to go beyond the argument that the problems are merely a *simple* expression of political ideology. In addition, the very urgency of some environmental problems – in particular the short time limits within which it has been estimated that certain scarce resources are likely to run out if they continue to be used at their present rate – has added an intensity to the whole debate on the nature of technology and made its implications very much more immediate.

Damage to the environment has always been associated with social and economic development. Long before the birth of Christ vast forests in the Near East were reduced to open plain by man's agricultural activities, and the loss of wind protection, together with centuries of over-grazing, soon turned these into arid desert. The problems of smoke pollution were so bad in medieval London that smoke control restrictions on log fires had eventually to be introduced in the seventeenth century and so great was the demand for wood from the forests of southern England during the sixteenth and seventeenth centuries that its use for burning as charcoal to smelt iron ore was banned to ensure a ready supply of material for ship-building.

In the past these types of environmental problem tend to have been relatively localized. A crucial aspect of present problems however is their global character. When a country's energy or mineral reserves run out, it can usually obtain further supplies from abroad; when the world's supply runs out, the only alternatives are to find a substitute, or do without. Already the United States is facing a growing energy crisis which has led to petrol rationing and official requests to conserve electric power. Most Americans had got used to the idea that there would be a fuel and power crisis some day, but the general expectation was that the critical shortages would not occur until the next century. Reports in 1973 that there were barely enough natural gas resources left to assure heating for homes, and that oil supplies in some areas were already running out, raised considerable alarm.

Programmes for replacing fossil fuels by the use of nuclear energy have run into both technical problems and heated opposition from environmentalists. It is now doubtful that nuclear energy or, indeed, any large-scale energy source able to replace fossil fuels, will arrive soon enough, in sufficient quantities and with its socially-hazardous effects adequately contained to prevent a rapid rise in the cost of energy as supplies of conventional fuels become shorter. The political and economic problems this could present are enormous. Between 1960 and 1970 the amount of energy imported into the United Kingdom – mainly as oil – rose from 25 to 45 per cent of total consumption (although temporary respite may come from the discoveries of gas and oil in the North Sea). The dependence of the USA on imported energy rose from 6 to 10 per cent over this period despite strong emphasis on attempts at self-sufficiency. And the Western world as a whole had a taste of possible political problems to come during the autumn of 1973, when the use of oil supplies as a political weapon by the Arab States, following the Middle East War with Israel, led to large-scale reduction in the supplies and the introduction by governments of wide-scale measures designed to reduce petrol consumption, as well as a sharp rise in the cost of oil.

The so-called 'energy crisis' is closely associated with the rapidly escalating use of energy in all branches of technological activity, both industrial and otherwise. Aluminium, to take a typical example, is one of the most widely used materials in modern industry. World supplies of aluminium are plentiful, with little danger of being exhausted in the near future. The main problem with the use of aluminium lies in the large amounts of energy required during its industrial processing. According to Professor Barry Commoner, aluminium requires about fifteen times more fuel energy per unit of 'economic good' than steel, and about 150 times more than timber, yet it has increasingly displaced both of these as a construction material. Twenty-eight per cent of all industrial use of electric power in the US is claimed to be accounted for by aluminium and chemical production alone.

To many, advanced technology appears to be the prime cause of our present environmental problems. This case is argued strongly by Professor Commoner in his book *The Closing Circle*. Commoner claims that the levels of different pollutants in the United States rose by between 200 and 2,000 per cent from 1946 to 1971, and that neither increasing population, nor a rise in the general standard of living, were sufficient to account for the most excessive of these increases. A survey of the average annual percentage change in production of several hundred items, which together were felt to represent a major part of the total US agricultural and industrial production, revealed some dramatic figures. The greatest growth was recorded in the production of non-returnable soda bottles, which Commoner claims increased by the equivalent of 53,000 per cent over this period. Second place went to the production of synthetic fibres, with a 5,980 per cent increase; mercury used for chlorine production increased by 3,930 per cent, the production of air compressor units by 2,850 per cent, plastics by 1,960 per cent, synthetic organic chemicals by 950 per cent, and electric power by 530 per cent.

These increases are far in excess of any increase in population or general affluence. To Commoner, they demonstrate clearly that the most drastic changes over this period were in the *type* of technology and production processes used by US industry, rather than changes in 'an overall output of the economic good', or any indications of the material aspects of consumption. Quoting examples such as innovations in agricultural technology, and the increasing use of plastics and non-degradable synthetic detergents, he concludes that 'the new technology is an economic success, but only because it is an ecological failure'.[2] He goes on to state his belief that 'the chief reason for the environmental crisis that has engulfed the United States in recent years is the sweeping transformation of productive technology since World War II . . . productive technologies with intense impacts on the environment have displaced less destructive ones. The environmental crisis is the inevitable result of this counter-ecological pattern of growth'.[3] Commoner's conclusion is clear. 'We are concerned not with some fault in technology, which is only coincident to its value, but with a failure that results from its basic *success* in industrial and agricultural production. If the ecological failure of modern technology is due to its success in accomplishing what it sets out to do, then the fault lies in its *aims*.'[4]

Various environmentalists have disputed Commoner's conclusion about the *primary* importance of productive technology as a cause of the environmental crisis. It has been pointed out that there are a number of individual cases in which the combined effects of population increase and greater affluence are sufficient to account for the major part of increases in the presence of particular pollutants. John Holdren and Paul Ehrlich, for example, writing in *Science and Public Affairs*, claim this to be true for nitrogen oxide emitted by automobiles. They argue against Commoner that 'faulty technology should be attacked directly wherever one finds it – a point we have never disputed. But the complacency Commoner encourages concerning population growth and rising affluence can only guarantee that the many environmental problems *not* primarily due to faulty technology will continue to worsen'.[5]

It seems generally agreed among environmentalists, however, that even if advanced technology is not the only culprit, it must nevertheless share a large part of the blame for the extent to which man is polluting the natural environment and using up the world's resources. Economists argue that the problem arises because industry has been able to ignore the so-called 'external costs' of production. These are the social costs imposed on the community by a particular production technique or technological development which do not enter into the usual accounting procedures. Pollution – the discharge of toxic effluent, for example, into a local river – is such an external cost. The elimination of pollution is offered, but at a price; and despite claims that 'the polluter must pay', the costs are usually passed on to the consumer through increased prices or taxation.

Despite fierce controversy over the methods used to predict the likely effects of pollution, overpopulation and resource shortage in the future, it is being increasingly brought home to us that there are very real finite limits to material consumption as well as to the degree to which the environment is capable of de-polluting itself. Some of these may indeed be closer than many of us would like to admit, and the important question now appears to be not whether it will be necessary to move to a steady-state economy, but when and how. Yet the existence of finite limits to resources is a fact which classical economic theory, despite its concern with concepts of scarcity, is virtually unable to take into account, at least when applied to the global situation. John Maynard Keynes even suggested the necessity of accepting a certain unavoidable level of pollution by predicting that 'for at least another hundred years we must pretend that fair is foul and foul is fair; for foul is useful and fair is not'.[6] But we no longer have a hundred years to spare. It is physically impossible for the world to continue tracing out current patterns of industrial and technological growth for this length of time, without taking into account the physical limits imposed by the finite nature of the earth and its resources.

A particular formulation of the case against contemporary technology is contained in the general critique of economic growth. The relationship between technology and economic growth is a complex one. It is argued by some that general scientific and technological development is itself sufficient to promote economic growth. According to this view, heavy support of research and development activities is one of the most effective ways of stimulating such growth.[7] However this process also works the other way. An economy based on rapid growth demands a particular type of technology, one which is capable of producing the maximum innovation of new products, often with little consideration of real social need. As one US businessman replied to a recent survey on industrial research policies, 'the problem will be to obsolete our products before someone or something else does'. The basic objective of industrial research is, in the words of a major Italian chemical company, 'the transformation of an idea or intuition into a new, improved or less costly product', rather than the immediate desire to tackle urgent social problems.[8] Given this symbiotic relationship, it is almost impossible to establish the exact nature of the links between the support of both public and private R & D activities and the rate of economic growth of a particular country (beyond the observable fact that the higher the level of industrialization, the greater is usually the proportion of gross national product spent on R & D). We appear to have an integrated situation; the two factors – economic growth and technological change – are so intertwined that it is impossible to talk merely in terms of the effect of one on the other. Policies requiring economic expansion have been reflected in the particular forms of technology through which this has been achieved.

We cannot therefore discuss the possible necessity of limiting – or even stabilizing – economic growth without paying attention to the technological factors that are inevitably involved. If a massive wastage of resources is incurred by the products of a technology economically-based on the need for innovation and obsolescence, any change to a resource-conserving philosophy requires both a technology and economy based on very different principles. For a society which has almost adopted economic growth as a policy-determining ideology, such a change is unlikely to be welcomed by those whose wealth relies on the interest that growth produces on capital investment, or indeed easily accommodated into the framework of classical economics.

If innovation and commercial competitiveness provide the incentive for the development of industrial technology, expenditure on research and development in the government sector is dominated by defence and military requirements. In Britain, these take up almost half of the total amount of money spent by the Government on R & D, accounting for £260m. out of £583m. ($1,327m.) in 1969. The military R & D budget in the US, at roughly $8,000m. in 1969, was almost as much as the *total* private and government expenditure on all branches of R & D in the United Kingdom, France and the Federal Republic of Germany combined. Technology has always been heavily dependent on financial support from the military sector. Many of the earliest engineers, Leonardo da Vinci being but one example, made much of their living from carrying out military projects for political leaders. The situation remains much the same today. In 1972 the increase requested by the Nixon administration in Federal support for military R & D projects was greater than the increase requested for all other areas of R & D – including health, transport, and social welfare – put together; 52 per cent of the R & D scientists and engineers employed by the US Federal Government in 1969 were working for the Department of Defense. A consequence of this concentration on military applications of technology in the US has been the build-up of a vast industrial-military complex. Major industrial firms – Honeywell, Lockheed, and so on – are now involved in the productions of machines whose main purpose is the destruction of human life. One does not have to be a pacifist to be appalled by the type of atrocities made possible by the techniques of 'scientific' warfare, with its present atomic, biological, chemical, electronic and even psychological dimensions.[9]

A new set of problems emerges when we look at the relationship of advanced technology to the problems of underdeveloped countries. Many development economists still support the idea that the key to development is to be found through sustained economic growth measured by aggregate factors such as gross national product. Growth is usually to be achieved through a concentration on industrial production. This is considered the first step away from a dependence on the export to the developed countries of primary commodities – tea, sugar, etc. – and an equally burdensome dependence on the import from these countries of manufactured goods. Such theories of growth and 'primary import substitution' legitimate the massive transfer of advanced industrialized techniques and private investment from the developed to the underdeveloped countries. They lead to an idea of development expressed in overall measures of economic performance, in particular stressing the importance of economic growth as a means of achieving a strong position in world trade. A report prepared under the Canadian ex-Prime Minister Lester Pearson for the International Bank for Reconstruction and Development in 1969 states unequivocally that 'the majority view of those who now administer aid, whether bilateral or multilateral . . . (is) that increased allocations of aid should be primarily related to performance'.

The introduction of advanced technology into underdeveloped countries, however, can bring with it many problems. The first is the drain it creates on foreign reserves, and the accompanying need for massive investment by outside interests. These reserves are used to buy the capital equipment required for advanced industrialized production, and often to buy in addition both the expertise and raw materials needed to operate the equipment. A related problem is the increasing technical dependence of under-developed countries on the developed countries, with important economic and political implications. Many multi-national companies prefer to set up their own subsidiary in a foreign country rather than hand over technical information to potential competitors through patents and licensing agreements. Often those agreements that are made have so many strings attached covering aspects such as marketing and further innovation that, as one Latin American economist has put it, 'the only decision left to the licensee is whether or not to enter into an agreement for the purchase of a technology'.

A third problem of rapid industrialization is that of unemployment. This is often associated with the substitution of capital-intensive industrial techniques for the traditional labour-intensive techniques required by craft-work and other forms of small-scale production. It also reflects migration of rural populations off the land into the towns in search of jobs. Unemployment is now estimated to be as high as 30 per cent in many underdeveloped countries. The lack of jobs partially results from the introduction into these countries of advanced technologies inappropriate to the prevailing social and economic conditions. As Professor Hans Singer, director of the Institute of Development Studies at the University of Sussex, has pointed out, 'the underdeveloped countries are now importing what is called modern technology, though the word modern, like the word progress, means what is modern, what is progress, for the richer countries'.[10]

The effects of advanced technology, compounded with a world trade system dominated by the developed nations, have contributed to the emergence of a double development gap. The first is between the rich and the poor countries. According to the World Bank, the advanced capitalist countries' share of world trade rose from 60 to 72 per cent between 1950 and 1969, while that of the 'less-developed countries' fell from 32 to 18 per cent over this period.

The second development gap is that which has occurred within the developing countries themselves. Over 80 per cent of those living in such countries still work on the land, yet technological innovation remains concentrated on the industrial sector, or on the intensive mechanization of agricultural processes. The main benefits from both of these accrue to a minute section of the population. 'The contrasts of wealth and poverty in many developing countries probably exceed those of any other civilization at any other time in history,' according to Judith Hart, a former UK Minister of Overseas Development.[11] Inequality is perpetuated by the availability and use of technology, a direct reflection of economists' emphasis on economic performance, rather than on the satisfaction of direct social need and the corresponding issues of political control.

To build a complete critique of contemporary technology, we have to go beyond purely environmental, economic and development considerations. Equally important are the effects that technology has on the individual in society. Even pollution problems can only be evaluated by the extent to which they affect the individual. This happens either directly, in terms of hazards to health and general well-being, or indirectly through depriving him of important material resources. Similarly an analysis of social priorities can lead only to directly political conclusions that embrace one's ideas of the position and role of the individual in society.

One of the apparent consequences of living in an industrialized society is that the individual finds himself further and further removed from many of the major decisions taken by – or perhaps one should say on behalf of – the society in which he lives. Even a leader in *The Times* (27 May 1968) discussing a speech in 1968 by the then Minister of Technology, Anthony Wedgwood Benn, on the need to increase participation in decision-making, commented that 'a paradox of modern technological society is revealed; the society creates problems so complex that they can be handled only by those with specialist skill and intricate knowledge, and at the same time it produces people who are in general more highly educated and inquiring than previous generations. It centralizes decision making but spreads the desire to make decisions'.

The spread of 'self-management' ideas, following in particular the success of work-ins at Upper Clyde Shipbuilders and elsewhere, demonstrate that this desire is by no means confined to the highly educated. Major decisions about a particular technological development, however, are taken not in the light of public debate about its general social desirability, but on its commercial viability and financial prospects in a world market. The debacle over the continuously escalating costs of the Concorde project is one example. Two other important, but less publicized, cases are the contracts to supply new telephone equipment to the British Post Office, and Britain's attempts to create a commercially viable nuclear power industry. In both of these cases, behind-the-scenes discussions involving many millions of pounds of public funds have been masked by a wall of technical argument. This argument is often used to deflect any 'non-expert' questioning – in other words to discourage direct public involvement. Any attempt at such intervention is often dismissed on the grounds that 'only the experts really know what they are talking about'.[12] Dr Harold Agnew, director of the Los Alamos Laboratories Weapons Division in the US, has been quoted as saying that 'the basis of advanced technology is innovation, and nothing is more stifling to innovation than seeing one's product not used or ruled out of consideration on flimsy premises involving world public opinion'.[13]

The political problems that result from the technical necessity of using a high level of expertise have been widely discussed. In his book *The New Priesthood*, for example, Ralph Lapp discusses the power and influence wielded by important scientists and technologists in the US. He quotes a remark made by Woodrow Wilson during his 1912 presidential campaign: 'What I fear is a government of experts. What are we for if we are to be scientifically taken care of by a small number of gentlemen who are the only men who understand the job? Because if we do not understand the job, then we are not free people.'[14] According to Ralph Lapp, the process has already happened, and democracy now faces its most severe test in preserving its traditions in an age of scientific revolution during which 'the nation becomes more and more dependent for its welfare upon the even fewer people who form a scientific and technical elite'.[15]

The extent to which the resources now required to service and maintain our technological society are welded into one mammoth 'technostructure' have been comprehensively mapped by economists such as J. K. Galbraith, and need little expansion here.[16] Scientists and technologists become increasingly linked with the mechanisms of power. Knowledge of, and hence advice on, scientific and technological issues becomes an essential part of the political process. But whether in fact scientists and technologists really wield the *power* feared by Woodrow Wilson and described by Lapp is open to question. Undoubtedly many are in a position to advise on decisions with wide social and political implications. At the same time, however, they are only given the freedom to come to those decisions which are or can be politically sanctioned.

A typical example of the need for political acceptability is the British Government's decision to ignore the conclusions of the Roskill Commission on the siting of London's third airport, and to accept an alternative site to that recommended by the Commission. Another is the fact that when the British Cabinet's 'think-tank' proposed to the Conservative Government in 1970 that the Concorde project should be abandoned, it is said to have been told, in the light of delicate Common Market negotiations, to think again; the 'think-tank' subsequently came back with the conclusion that Concorde should press ahead, and Anglo-French relations were preserved. Political considerations were similarly involved in the Government's decision – based on the recommendations of Lord Rothschild – that a large part of the activities of the scientific research councils should be determined by the requirements of government ministries and departments. While publicly justified in terms of greater 'efficiency' and 'accountability', this restructuring reflected a desire to strengthen and institutionalize the links of the technostructure in which government, industry and the universities combine to form a large corporate consensus. In the US, those scientists and technologists prepared to place their skill at the disposal of the American war effort in Vietnam were rapidly absorbed into the military establishment. This was done through, for example, the Jason group of the Institute for Defense Analysis, a committee made up of an élite selection of about 40 academic scientists from various US universities. Equally eminent scientists and technologists who spoke out forcefully *against* the war in Vietnam, producing documented evidence of the massive ecological damage caused by the use of herbicides and defoliants, were relatively ignored by those responsible for US military policy.[17] Such examples indicate clearly that scientists and technologists wield power only through their adhesion and allegiance to an existing political base.

We should also be concerned about the fact that scientists and technologists are frequently used, not merely to produce the instruments of political action, but also to add a mask of objectivity – and hence political neutrality – to political decisions. The presentation of political issues as technical ones, accessible only to 'expert' debate, has important implications. It legitimates the individual having less and less say in major decisions – which may range from the future development of the telephone system, or of supersonic transport planes, to Britain's joining the Common Market, a political decision 'justified' by economic as well as cultural arguments – that are likely to have an important effect on his life. Often the only way to challenge such decisions is by collective pressure-group activity. Here it is those with the largest amount of resources at their disposal that have the greatest chance of success; expert advice, unless offered voluntarily, can be expensive, and is thus usually available only to those who can afford it.

Taken to its extreme, the 'depoliticization' of advanced industrialized society could have even more severe consequences. It provides a powerful weapon for achieving the closely co-ordinated control of the actions of all members of society – the path towards totalitarianism. This tendency has been made increasingly possible by developments in cybernetics – the science of controlling information and communication – and in the design of computerized communication networks. Early in 1973 Chile was conducting experiments under the guidance of a British cybernetician, Stafford Beer, to find the extent to which the country's economy could be regulated through the use of a computer to record the daily activities of every factory. It was also claimed to be trying to discover whether it is possible to predict the effects of political and economic measures.[18] Echoes of totalitarianism are already contained in the report of the Royal Commission on Local Government in England, the Redcliffe-Maud Report, published in 1969, which states that 'the new and more sophisticated techniques of management also give an impetus to the development of central management. Fresh prospects are opened up by the computer'.[19]

Professor Frank George, director of the Institute of Cybernetics at Brunel University, has argued that 'the real danger that cybernetics . . . may bring about, and indeed already is, is that it makes more likely and more easy the establishment of a fascist autocracy: a community wholly controlled by central government'.[20] A study of totalitarian dictatorship and autocracy by two social scientists, Carl Friedrich and Zbigniew Brzezinski, reached the conclusion that four out of six elements common to most dictatorships and autocracies – namely monopolistic control of communications, weapons, terror and the economy – relied on some form of technological base.[21]

This is not to imply that the increasing use and reliance on technology or the application of cybernetics to social planning itself leads to totalitarianism. Indeed, dictatorships have existed at all levels of technological development. The main point, however, is that the increasing 'depoliticization' associated with advanced technology could make such a system appear a 'rational' economic necessity lying outside the field of political considerations, a scenario predicted by George Orwell in his *Nineteen Eighty-Four*. This tendency is already apparent in various arguments put forward in support of highly authoritarian policies for population control. By dressing up political issues as complex technical ones, the technocracy removes responsibility from the individual and by doing so, appears to isolate him from the political process.

Already we can see how contemporary technology tends to accentuate and reinforce class divisions and inequality rather than remove them. Those who lack sufficient resources to make use of available technology are placed at a permanent disadvantage. We see, for example, the plight of those who have been labelled the 'transport poor', members of the community who are penalized and discriminated against for not owning a private car by the lack of adequate public transport facilities. Similarly many of the benefits of contemporary medical research are devalued in Britain by a National Health Service which permits those patients who can afford private treatment almost immediate access to a consultant's skill and training, while others may have to wait months or even years for such treatment. New telephones can be provided immediately for British people requiring them for business or commercial purposes; pensioners and others in need often have to wait years to be connected.

Far from being the great democratizer that was to bring equality to all, technology has become yet one more means by which one social class maintains supremacy over another. The more sophisticated that technology becomes, the greater is this effect. The British Society for Social Responsibility in Science points out in its policy statement that 'Scientific and technical knowledge is an important source of power, but only large institutions have the resources needed to exploit it . . . thus science is used directly to increase the power of the already powerful and to frustrate the expectations of the powerless'.[22]

As well as removing him from the political process, technology also provides the means by which the worker is removed from control over the process of industrial production and presents him with the pre-defined role he is required to play in this process. Karl Marx used the term *alienation* to describe the way in which, under the capitalist system of production, the product of a man's work is immediately taken away from him, and he loses all control over what happens to it. To Marx, productive activity was the main way in which man established his relationships both with the natural world, and with the other members of society. It was also the process by which the individual fulfilled his full potential as a human being. Marx saw work as the concrete manifestation of that which differentiated man from the rest of the animal kingdom. In *The German Ideology*, for example, he writes that men 'begin to distinguish themselves from animals as soon as they begin to *produce* their means of subsistence, a step which is conditioned by their physical organization'.[23]

Marx identified four aspects of alienation in capitalist society that impeded this process: man's alienation from his work, from the product of his work, from other men and from the human species in general. Alienation was primarily a function of the relations of production rather than of the particular modes of production (the machines and techniques used by a society); it remained specific to capitalist society, and would disappear once this had been substituted by a communist regime.

The concept of alienation has been widely debated since Marx's time. Some dismiss it as an 'unscientific' concept belonging to Marx's early, more 'humanist', writings; they point to the fact that it is scarcely referred to in the three volumes of *Capital* to argue that by the time Marx had come to write his major work, he had 'corrected' his earlier ideas and had discarded alienation as a useful concept.[24] Others see this as a misinterpretation of Marxism, and have developed Marx's idea of alienation further, particularly in the light of the theories of Freud and others concerning the nature of the human unconscious. Some of these have attempted to extend Marx's critique to include the idea that man is alienated from himself – or 'dehumanized' – as much by the nature of the contemporary social environment as by the fact that it is controlled by a dominant social class; they suggest that the modes of production may be as important as the relations of production in determining man's state of alienation in industrial society. The psychologist Erich Fromm, for example, has written that 'man, as a cog in the production machine, becomes a thing, and ceases to be human . . . the passiveness of man in industrial society today is one of his most characteristic and pathological features . . . being passive, he feels powerless, lonely and anxious'.[25] Fromm refers to this as the 'syndrome of alienation'.

The somewhat abstract nature of Marx's concept of alienation makes it difficult to apply as an analytical tool to particular work situations. Five different aspects of alienation, however, utilized by sociologists and psychologists as operational categories have been identified by Melvin Seeman. These roughly correspond to Marx's own four aspects of alienation. They are: the *powerlessness* of the individual when he feels himself controlled or manipulated by other people or by an impersonal system (such as technology); the *meaninglessness* that he associates with his work, a feeling increased by the division and fragmentation of production tasks, and often apparently encouraged by bureaucratic structure; the *self-estrangement* of the worker who often experiences a form of depersonalized detachment from his work; and finally a general *normlessness* and *isolation* that can lead to social alienation or the general break-up of integrated communities and is often associated with the French sociologist Emile Durkheim's concept of *anomie*.[26] These four aspects of alienation were used by the US sociologist Robert Blauner in a wide-ranging study of the factory worker in American industry. His general conclusion was that, although each dimension of alienation varied in form and intensity according to the industrial system being studied, 'inherent in the techniques of modern manufacturing and the principles of bureaucratic industrial organization are general alienating tendencies.'[27]

Blauner found that workers in an automated chemical plant showed a far lower degree of subjective alienation than those working in a textile factory or on an automobile production line. He ascribed this to factors such as the unpredictability of chemical processes which adds 'interest, excitement and challenge', the flexibility in work rhythms that is made possible by automatic control, and the increased amount of 'on-the-job leisure' that was usually taken up with 'reading . . . conversation and joking with fellow workers'. Blauner's general conclusion from this was that 'since work in continuous-process industries involves control, meaning and social integration, it tends to be self-actualizing instead of self-estranging'.[28] In other words, increased automation and the work-style that this introduces is seen by Blauner (and a large number of liberal industrial sociologists) as one way of removing the problem of alienation.

What Blauner's analysis of automated work was unable to foresee were the changes that have been brought about by the development of more sophisticated control techniques in the ten years since his studies were carried out. Fail-safe and other automatic monitoring devices have tended to reduce the involvement of the operator to a minimum, once again removing much of his regained sense of responsibility. In such circumstances, alienation and frustration are only aggravated by the process of automation.[29] One survey of automobile workers found that many complained of the social alienation brought about for example by the greater distances between work posts in the automated workshop and by the fewer breaks and opportunities for contact between workers. They complained of a feeling of estrangement from their productive process under automated work conditions.[30] According to one sociologist, B. Karsh, 'automation seems to disconnect man from the machines . . . mass production having dispossessed the man of the creative forms of work, he remains dispossessed by automation of control over his work.'[31]

A major effect of the introduction of increasingly capital-intensive technology results from the requirements it places on the habits of the workforce. Above a certain level of capital investment, a maximum return can only be achieved through the maximum use of equipment, and labour charges become a relatively insignificant part of the overall costs. A survey of British firms carried out by management consultants concluded that 'the average British company's computer is used only half the time it could be, and is productive for only two thirds of the time it is used, *wasting* about £3,000 a month'.[32] The total wastage bill for 104 computers investigated exceeded £3½m. a year, this situation being the result, according to a director of the firm which carried out the survey, of 'serious management failings'. To avoid this type of 'wastage' computer operators are now obliged to organize their lives to meet the demands of the computer, and not the other way round. According to one British trade unionist, 'high capital equipment, such as computer aided design equipment, is now becoming more widespread in technical areas. In consequence of that, the employers will wish to ensure that all employees who use this equipment accept the same type of subordination to the machine that they have already established on the workshop floor'.[33] A direct result is the relatively new phenomenon of university graduates finding themselves on shift-work, and the general 'proletarianization' of technical workers.

The problems of unemployment and redundancy created by automation are more difficult ones to discuss. Undoubtedly the increasing productivity of British industry in recent years has been achieved at the expense of a declining number of jobs in particular industries or firms. In GEC, for example, the largest private employer in the UK, the number of employees fell from 230,000 to 181,000 between 1969 and 1972, while profits rose over this period from £49m to £77m. Undoubtedly, too, it is possible to correlate redundancies in certain firms and industries with the introduction of certain types of automation. One can point to the construction of vast chemical complexes representing an investment of many million pounds, yet employing no more than a dozen people.

The reasons for the dramatically high unemployment figures in the early 1970s however – including for the first time a high level of unemployment among qualified scientists and engineers – were as much the result of political and economic policy as they were due to changes in the technology of production.[34] Indeed, to blame such unemployment *directly* on the effects of automation is yet another example of 'depoliticization', the concept of 'efficiency' being invoked to justify measures taken on the grounds of economic performance rather than social need. Instead, we should look for the causes in the functioning of a system which prefers to employ many thousand more skilled engineers and draughtsmen on the design of supersonic aircraft, or sophisticated military technology, than on the problems of urban public transport systems or medical equipment.

Taken individually, a large number of the social problems associated with advanced technology that we have discussed may well have piece-meal technical solutions. These often lead politicians and others to affirm a strong belief that such problems can be solved by new and better technology, in the same way that they suggest that the problems associated with economic growth can only be solved through further economic growth. They point out that many pollution problems, for example, can be solved or diminished by the introduction of adequate pollution-control devices, and occasional adaptations to production processes. Similarly, some of the more immediate psychological problems created by the stress of modern working conditions can be relieved by the introduction of techniques such as 'job enlargement' and the re-integration of previously fragmented work-tasks. The problems of city transport could (theoretically, at least) be solved by the elimination of the private motor car and a massive financial and technological investment in computer-controlled overhead transport networks. The list of such 'technical fixes' that are available could be extended indefinitely, for the problems which technology is best able to solve are precisely those which have been isolated from their social environment. Faith in technology is left undented. Referring to the problems of pollution and nuclear warfare, Donald Schon has written that 'while these unwanted effects of technology are disturbing, they do not leave us without a clear course of action. We must simply apply more energetically and more exhaustively the kind of practical reason central to techno-logical advance'.[35] A more sophisticated approach to the 'technical fix' is the development of social institutions designed to predict the likely social consequences of particularly technological develop-ments (such as the recently established Office of Technology Assessment in the US) so that attempts can be made to minimize these before they occur.

Despite such frequent reaffirmations of faith in the powers of technology by both politicians and technologists, attitudes towards science and technology have begun to change dramatically over the past few years. The quarter of a century that followed the Second World War, and particularly the 50s and early 60s, was a boom-period for technology. Scientists and technologists, having won their spurs on successful wartime projects such as the develop-ment of radar and the atomic bomb, found themselves riding the crest of a wave of optimism. Spending on research and develop-ment in the US increased almost 60 times between 1940 and 1965, from $340 million to $20,000 million a year. Similar increases occurred in most other advanced industrial countries. Political leaders went round publicly affirming their confidence in science and technology to come up with the solutions to all the world's major problems. President Kennedy, for example, announced in 1963 that 'as we begin to master the potentialities of modern science we move towards an era in which science can fulfil its creative promise and help bring into existence the happiest society the world has ever known'. Mr Harold Wilson described the Labour Party's aim in 1960 as being to harness socialism to science and science to socialism; four years later, a few months before becoming Prime Minister he was telling a conference in Birmingham that Britain's future was 'to be forged in the white heat' of a government-inspired technological revolution.

By the mid-60s, however, the wave of optimism had begun to break. Once the race to get a man on to the Moon had been won, there was an immediate decrease in expenditure on space-related research, which fell from a total of 20.4 per cent of the US Federal Government's total R & D expenditure in 1965 to almost half this figure in 1970. The cut-back in space research contributed to a 16 per cent decrease in the number of research scientists and engineers employed in the aerospace industries between 1970 and 1971. The failure of the US to achieve any form of victory in the Vietnam War, despite its vast technological supremacy, severely dented the official success image of technology even further. In Western Europe, the apparent inability of technology to help solve the economic problems experienced by many countries in the late 60s lead to a gradual dampening of political enthusiasm for spending large sums of money on research and development with no immediate economic pay-off in sight. While General de Gaulle had previously emphasized the need to 'push relentlessly our technical and scientific research in order to avoid sinking into a bitter mediocrity or being colonized by the activities, inventions and capacities of other countries', many politicians now seemed to agree with the Treasury official, giving evidence before the House of Commons Expenditure Committee in its 1971/72 session, who remarked that 'most people would feel that the return the nation has had from its R & D expenditure has been disappointing'.

The path from disillusion to distrust is a short one. In January 1973, Mr Harold Wilson gave a speech in Edinburgh warning of the dangers of technological advance, and describing the forces of technology which were 'riding across people's lives', leaving a trail of misery and destruction. The problems of man in modern technological society, he said, his frustration, his tension, his seeming impotence, were common to every advanced country. The contrast with his speeches of ten years previously, with their high gloss of technological euphoria, could hardly be more marked. The next issue of *Private Eye* contained a cartoon in which Mr Wilson was instructing his secretary to 'dig out my election speeches of 1964 and replace wonders of technology with horrors of'.

This disillusion has been reflected and sustained by several writers who have attempted to synthesize an analysis of the problems of our modern technological society. The most prominent of these have been Jacques Ellul, Herbert Marcuse, Jürgen Habermas and Theodore Roszak. Ellul's main attack, contained in his book *The Technological Society* published over twenty years ago, has been directed at *la technique*. This is perhaps best translated as technological practice, which he says has 'fashioned an omnivorous world which obeys its own laws and which has renounced all tradition . . . man himself is overpowered by *la technique* and becomes its object'.[36] His arguments bear close resemblance to those of the Victorian writer Samuel Butler in *Erewhon*, whose central character comes across a society that had banished the machine, after it had realized the increasing oppression and dominance of man by 'mechanical consciousness'. Marcuse's *One-Dimensional Man* and Roszak's *The Makings of a Counter Culture* are both strong attacks on industrial society. For Marcuse, 'the liberating force of technology – the instrumentalization of things – turns into a fetter of liberation: the instrumentalization of man',[37] while Roszak states that 'in the case of technocracy, totalitarianism is perfected because its techniques become progressively more subliminal'.[38]

Until recently, most of these attacks had fallen outside the mainstream of political critique. They could be dismissed by both Left and Right as little more than the rumblings of discontented or frustrated romantics. To question the very nature of technology, rather than merely the uses to which it was put, seemed to challenge the whole basis of rational thought; the apparent political neutrality of both science and technology left responsibility for these activities as part of the conventional political process. Attacks on the 'counter-culture' philosophy in particular, with its almost complete rejection of the conventional life-style of contemporary industrial society, have tended to look upon the move away from modern science and technology as some form of regression to the dark ages. Jacob Bronowski has described the counter-culture as 'a recapitulation in modern dress of the anti-intellectual, irrational and illiberal prejudices that have always been endemic in America. An armoury of old and scaly prejudices is being foisted on the young in the disguise of a gospel of nature'.[39]

Whatever one feels about the political significance of the counter-culture movement, one of the more important practical ideas to have emerged from it is that which forms a major theme of this book, the need to develop an alternative technology. The roots of alternative technology are to be found equally in the social and political critiques of those concerned with what they see as the anti-human and alienating aspects of contemporary technology, and among those who argue primarily on environmental grounds that the polluting and resource-wasteful elements of such technology make the search for an alternative mode of technological development an urgent necessity.

Alternative technology is now attracting a growing number of advocates from each of these camps. Its general approach starts from the premise that the roots of the problems created by modern technology are to be found as much in the design of the technology itself as in the uses to which it is put. Eschewing any patching-up attempts of piecemeal social engineering, it suggests that solutions can only be found through a radical overhaul of society's technological and industrial base. Robin Clarke, one of the co-founders of Biotechnic Research and Development (BRAD), has described 'the new view' as being 'that science and technology will not themselves find a way out of the present crisis – but that any real way out will involve a science and a technology, even if those activities in the future bear little qualitative or quantitative resemblance to science and technology today'.[40]

It would be wrong to imply a complete harmony of interests among those who make up what might loosely be called the alternative technology movement. To a certain extent the different approaches are reflected in the particular names that have been selected to describe the activities in question. Some of the names suggested, and adopted by different groups or individuals, include: soft technology, radical technology, low impact technology, intermediate technology (applied, in particular, to the technological requirements of the underdeveloped countries), people's technology, liberatory technology, and so on. The approach of each group however usually contains some combination of a set of common elements. These include the minimum use of non-renewable resources, minimum environmental interference, regional or sub-regional self-sufficiency, and the elimination of alienation and exploitation of individuals. Groups differ in the relative emphasis placed on these, however, and the differences between groups are often as marked as their similarities. Some consider alternative technology as a type of 'insurance policy' to cover the possible technological collapse of society through some massive ecological disaster. Others see it as a means of preventing such a collapse from occurring, and as forming the only sound basis for future social development, with each country developing a set of technologies relevant to its own particular resources and needs. While for many, alternative technology means the tools and machines by which both man and nature will be liberated from the domination and exploitation inherent in our present technology, emphasizing the extent to which it provides a framework for both a political and a technological alternative.

When we talk of alternative technology, therefore, we are primarily referring to a set of approaches to the alternative design and use of machines and tools, rather than to a particular set of machines. The concept of a commune-based society in which the activities of all communes are co-ordinated through a complex computer network may well qualify as an 'alternative' technology in terms of self-sufficiency and the decentralization of control: it might be argued, however, that the expertise and resources required to set up such a computer network would immediately exclude it from the alternative technology category. Whichever attitude is adopted, the main importance of alternative technologies does not lie in the particular solutions which may be offered to certain problems. Rather it is in the approach that they represent, that technology should be designed to meet human needs and resources – and not the other way round – and the recognition that radically different patterns of technological development may not only be desirable but necessary.

None of the social problems associated with contemporary technology, when taken in isolation, are sufficient to condemn the whole basis of modern technology. Looked at as a whole, however, it is difficult to avoid the conclusion that what is required is the development of an alternative technology which is based on political and environmental criteria which allow potential problems to be taken into account before the technology has been designed rather than after it has been implemented. Such an alternative technology might coincide with Aldous Huxley's suggestion that we need a 'differently orientated technological progress (resulting in) a progressive decentralization of population, of accessibility, of ownership of the means of production, of political and economic power'.[41] Equally it ties in with Gandhi's recommendation that 'every machine that helps every individual has a place, but there should be no place for machines that concentrate power in a few hands and turn the masses into mere machine-minders, if indeed they do not make them unemployed'.[42] Both underline the fact that alternative technology may prove in the long-run to provide the only basis for a life-style that is personally fulfilling, socially just and economically and ecologically viable.

2 The Ideology of Industrialization

How can we make sense of the various social problems associated with contemporary technology outlined in the previous chapter? In other words, how are these problems related to each other, and where do their root causes lie? To answer such questions, we must turn to the relationship between social and technological change. Social problems cannot be understood merely by a description and analysis of their contemporary manifestations. Their historical roots must also be examined. To explain why technology has had the effects that it has, it is necessary to go beyond a description of particular technological developments and their consequences. We must also look at the specific set of conditions that initially gave rise to these developments. By retracing the history of technology, some idea will begin to emerge of the incentives behind the process of technological innovation. These will indicate the type of social and political values implicit in the design of contemporary technology, beyond those of a primarily economic nature.

The traditional interpretation of technological innovation is that its prime intention is to increase the efficiency of industrial production. This is claimed to be done in order to maximize the production of surplus value (and hence, in capitalist societies, of profit). I hope to demonstrate in this and the following chapter that this interpretation is inadequate to explain what actually happened in particular historical situations. It is in fact an 'ideology', a particular interpretation placed on phenomena or events that distorts their essential nature in the interest of political ends.[1] In this case, the ideology has proved an extremely powerful tool in the history of social development. For it has promoted the idea that technology somehow possesses an internal objective logic that determines a unique progression from one stage of development to the next. Technology is therefore held to remain politically neutral in any particular situation, playing an entirely passive role with respect to issues of power and control. We shall call this interpretation the 'ideology of industrialization'.

The message of this ideology is that industrialization through technological development is a practical – rather than a political – necessity for achieving social development. It implies an objectivity to the process, and seeks to remove it from debate on political issues. It thus gives a legitimacy to policies that appear to promote the process of industrialization, often regardless of their political, or even social consequences. The social squalor experienced during the Industrial Revolution becomes justified as the 'price' that had to be paid for human progress. The same is said of the present destruction of both the natural environment, and of traditional human communities. To stand in the way of technology is, almost by definition, to be labelled reactionary. Industrialization is equated with modernization, with progress, with a better and healthier life for all.

The same formula is held out to the underdeveloped countries. They are told that their only hope for the future is to initiate the pattern of industrialization already followed by the developed countries. This is sold to the underdeveloped countries not on its political or social desirability, but as a technical necessity (whether politically desirable or not). Professor T. S. Ashton ends his book on the Industrial Revolution with the words: 'There are today on the plains of India and China men and women, plague-ridden and hungry, living lives little better, in outward appearance, than those of the cattle that toil with them by day and share their places of sleep at night. *Such Asiatic standards, and such unmechanized horrors, are the lot of those who increase their numbers without passing through an industrial revolution.*'[2]

Industrialization has undoubtedly succeeded in substantially raising the health and standard of living of a large number of the world's population. Its major achievements in these directions cannot be denied. But it is important to distinguish the essential character of the process of industrialization from the ideology to which it has given rise. What the ideology disguises is the degree of political exploitation and manipulation that has, in almost all cases, accompanied the industrialization process and hence the development of contemporary technology. Industrialization has appeared to necessitate, and has hence legitimated, man's exploitation of both fellow men and the natural environment. The apparent need for authoritarian discipline and hierarchical organization of the factory required to operate complex production-line equipment, for example, is held to justify the accompanying relationships between management and workers. Technological innovation seems to legitimate these relationships by increasing the degree of discipline required to run a factory 'efficiently'. According to Ashton, for example, 'the new methods of administration, the new incentives, and the "new discipline" were as much part of the [Industrial Revolution] as the technical inventions themselves: adaptation to them was the price the workers had to pay for the higher incomes that large-scale industry brought'.[3]

The technology of Western society as we shall see has come to embody the exploitation of both men and the natural environment which have characterized the political system within which it has been developed. According to André Gorz, 'productive forces are shaped by the capitalist relations of production . . . the imprint of the latter upon the first is so deep that any attempt to change the relations of production will be doomed unless a radical change is made in the very nature of the productive forces, and not only in the way in which and in the purpose for which they are used'.[4]

We shall explore further the political nature of technological innovation and its implications for advanced technology, in the following chapter. First, however, it is necessary to examine the various conventional interpretations of the history of technology contained within the value-free notion of industrialization.

The traditional model of social evolution is usually taken to be a straightforward pattern of linear development. The Stone Age, when man first learnt to use primitive tools and instruments, is considered the 'dawn' of civilization. After many thousands of years, this was followed by the Bronze Age. During this period man discovered the advantages to be gained from making his tools with metal, in particular its many agricultural and military uses. Bronze however is a relatively soft alloy. It was only the subsequent discovery and use of iron that provided civilization with the material base from which it could 'take off'. Iron ores were relatively abundant in the countries of the Eastern Mediterranean, and the spread of both smelting and working techniques made it the main metal of both Greek and Roman civilizations. Now frequently hardened through the addition of carbon to make steel, iron still forms the material backbone of all advanced societies. To a certain extent twentieth-century man can still consider himself to be an inhabitant of the Iron Age.

This interpretation of the history of civilization, with its unidimensional view of progress, implies that societies can be looked upon as primitive or advanced according to the level of their technological development. It formed the basis of almost all cultural and anthropological investigation up to the early years of the twentieth century, and is still the one most commonly used to indicate levels of 'development'. (It is also the most popular description of school text books, thus ensuring that the interpretation is maintained by the educational system.) The implied model of social evolution is frequently based on the concept of technological determinism, the idea that social development is almost entirely determined by the type of technology which a society invents, develops, or is introduced to. The anthropologist Leslie A. White, for example, has written that 'it is not changing sentiment that turns the wheels of social evolution. Rather it is the social and political groupings by the operation of technological forces that determine the direction and scope of the sentiment . . . the motive power of a culture, so to speak, lies in its technology, for it is here that energy is harnessed and put to work'.[5]

The implications of this concept are simple and attractive. Each generation throws up a few inventors (starting, perhaps, with the man who discovered fire). Their inventions appear to become both the determinants and the stepping stones of human development. Unsuccessful inventions are condemned by their failure; successful ones soon prove their value, and are rapidly integrated into the social environment, which they then proceed to transform. In this way, a technological 'break-through', in itself often held to be a fortuitous event, independent of environmental influences, is claimed to have had important social 'consequences', and one is permitted to study the 'impact' of technology on society. Thus the growth of early cities has been explained as a direct result of the development of agricultural techniques that enabled man to raise himself above a mere subsistence level and think in terms of trade and culture; the emergence of the Roman Empire has been attributed to the superiority of its military and communications technology; and the flowering of the Renaissance to the introduction of the printing press into Europe.

In a more detailed way, the historian Lynn White has traced the development of the feudal system in medieval Europe to the introduction of the stirrup. By increasing the control of a horseman over his mount in battle, White claims that the stirrup revolutionized military technology and the whole art of warfare. Once the horse had been brought to military prominence, he suggests that a sudden and urgent demand for cavalry led medieval Europe to reorganize its realms along feudal lines. This meant its rulers could obtain mounted troops in much greater numbers than ever before, each feudal village providing the resources for a single knight. White claims that 'feudal institutions, the knightly class and chivalric culture altered, waxed and waned; but for a thousand years they bore the marks of their birth from the new military technology of the eighth century'.[6] White also discusses the agricultural implications of the introduction of the heavy plough. This was 'an agricultural engine which substituted animal-power for human energy and time'. It not only changed man, according to White, from being a part of nature to being her exploiter, but also vastly increased the food-producing capacity of the countryside. Together with the development of new forms of food supply that included the use of vegetables such as peas and beans, he claims that this 'goes far to explaining, for Northern Europe, at least, the startling expansion of population, the growth and multiplication of cities, the rise in industrial production, the outreach of commerce, and the new exuberance of spirits that enlivened that age'.[7]

Many interpretations of the Industrial Revolution follow the same pattern. They claim it to be the direct result of major technological developments such as spinning machines, the use of coal and the introduction of steam-power. Even Frederick Engels, for example, before his contact with the economic theories of Marx, felt the Industrial Revolution to have resulted directly from improvements in technology. 'The invention of the steam-engine and of machinery for working cotton', he wrote in 1844, 'gave rise, as is well known, to an industrial revolution, a revolution which altered the whole civil society.'[8] Describing the social disruption that accompanied the technological innovations, he writes that 'the division of labour, the application of water and especially the application of machinery, are the three great levers by which manufacture since the middle of the last century, has been putting the world out of joint'.[9]

The idea that technology develops independently of society, and is subsequently imposed upon it, is still the most common model used today. Thus R. J. Forbes claims that 'technology . . . has become the prime source of material change and so determines the pattern of the total social fabric'.[10] One of the most outspoken proponents of this idea has been Marshall McLuhan, who attracted widespread interest in the 1960s with his ideas on the way in which technology had affected man's experience of himself and society. 'The medium is the message . . . the personal and social consequences of any medium – that is, of any extension of ourselves – result from the new scale that is introduced into our affairs by each extension of ourselves, or by any new technology',[11] he writes in *Understanding Media*. McLuhan takes up, for example, Lynn White's interpretation of the technological origins of the feudal system, and goes on to suggest that as a result of possible technological innovations such as shopping from home by 2-way television, 'if the motorist is technologically and economically far superior to the armoured knight, it may be that electric changes in technology are about to dismount him and return us to the pedestrian scale'.[12] Again, the emphasis on social change *determined by* apparently autonomous technological developments.

Theories of technological determinism have provided many valuable insights into the relationship between technological and social change. Unfortunately, however, there are a number of historical facts that cannot be interpreted in such simple terms. There have been many inventions that were ignored or even suppressed by the society in which they were first produced, some to be re-discovered or re-invented many centuries later under different social conditions. The use of steam to drive elementary machinery, for example, was known to the ancient Greeks, and the library at Alexandria contained a perfectly working model of a steam engine. Yet it was almost two thousand years later that the idea was put to a practical application. Although this was partly due to the fact that Greek society as a whole lacked the necessary technical and mechanical knowledge required to exploit sources of power on a large scale, it was also a result of the fact that a society such as ancient Greece had little need for major power-producing – or even small-scale labour-saving – devices. A plentiful supply of slave labour meant that the production of power presented few problems. The Emperor Vespasian, when presented with a design for a mechanical jig which he was told would do the work of many men, is said to have replied 'Take it away; I have my poor to feed'.

There have also been societies with technical skill and knowledge comparable to that which existed in Europe in the early Middle Ages, but which remained at a relatively low level of development. Professor Joseph Needham, for example, has produced considerable evidence to demonstrate the sophisticated technological knowledge of the Chinese. Not only were they the first to develop, among other things, paper, printing, the magnetic compass and gunpowder, but they also achieved impressive feats of engineering. These included the bridge built across the Chiao Shui River in southern Hopei, a graceful stone arch as shallowly curved as the Pont de la Concorde in Paris, and unmatched by any Western bridge for 500 years. Needham attributes the early history of clocks to China, where an enormous astronomical clock-tower was built in AD 1090. The Chinese were also using paddle-wheel vessels, driven by tread mills, long before the West. Yet China never developed its technology beyond this relatively early stage, and until the publication of the first volume of Needham's massive *Science and Civilisation in China*, had often been treated as technologically backward.[13] Similar traces of limited technological development can be found in, for example, India and Latin America.

In the light of such examples, it is no longer possible to stick to a simple model of technological determinism. A more sophisticated approach suggests a functional interdependence between technological developments on the one hand, and economic development on the other. According to this interpretation, the machines or techniques which have been developed at any one time in history were those most suited to the demands of the prevailing economic conditions. Similarly these economic conditions themselves encouraged the further development of 'appropriate' technology.

The combination of technological and economic development was felt sufficient to lead civilization along the path of industrialization and general progress. 'Human progress reflects directly the contributions of science and management to society,' according to an article in a management journal. 'Science created the knowledge and made it available for human progress . . . Management developed to utilize effectively the products of science for economic growth and social progress.'[14] This general interpretation acknowledges the fact that technological innovation is a social process. It also stresses the importance of economic factors. An innovation will only be introduced if it can be shown to be 'economically viable'; if not, however ingenious it may be, it stands little chance of social acceptance. Although this begs the question of how one evaluates the concept of 'economic viability', it is sufficient to explain why labour-saving devices were of little interest to the Greeks. It also shows how, in general, the lack of an adequate economic framework was sufficient to stifle technological development in many early civilizations.

This concept of functional interdependence between technological and economic development again represents the growth of technology as a linear evolution through successive stages along an inevitable course. At each stage of human history, technology and social organization are claimed to have progressed along independent but parallel paths. Primitive societies, according to this interpretation, possessed primitive technologies suited to rudimentary tasks, but seldom reaching any great degree of sophistication. Even in early 'civilized' societies, such as the Greeks and Romans, technology remained at a relatively simple level. The Middle Ages saw the first widescale use of elementary machine technology; there are said, for example, to have been 5,000 mills in England as recorded in the Domesday Book, and the mill formed an important part of the feudal village economy. Other areas in which elementary machines began to be used were weaving and agriculture. Most of the technology used, however, was still at the craft level. The individual craftsman was not only expected to have sufficient knowledge to mend or replace any part of his equipment when it broke, but frequently built the equipment himself in the first place. Craftsmen were brought together in guilds, which jealously guarded the secrets of a particular trade passed down from one generation to the next, and insisted on long periods of apprenticeship for those who wished to take up the craft.

In his book *Technics and Civilisation*, Lewis Mumford divides the history of technology into three successive but overlapping and interpenetrating phases. These are the eotechnic, the paleotechnic and the neotechnic phases, following a taxonomy first suggested by Patrick Geddes. Mumford labels the period between 1000 and about 1750 as the eotechnic phase. During this period, he writes, 'the dispersed technical advances and suggestions of other civilizations were brought together, and the process of invention and experimental adaptation went on at a slowly accelerating pace ... this complex reached its climax, technologically speaking, in the seventeenth century, with the foundation of experimental science, laid on the basis of mathematics, fine manipulation, accurate timing and exact measurement'.[15]

It was towards the end of the eotechnic phase that medieval craft technologies began to find themselves unable to cope with the growing demands of trade. These demands had been stimulated by a gradual increase in the overall standard of living and the opening up of foreign markets. The need for greater productivity led to the creation of workshops and the beginnings of the factory system of production. Marx and Engels gave a generally-accepted characterization of the situation in the *Communist Manifesto*. 'The feudal system of industry,' they wrote, 'in which industrial production was monopolized by closed guilds, now no longer sufficed for the growing wants of the new markets. The manufacturing system took its place. The guild-masters were pushed aside by the manufacturing middle-classes; division of labour between the different corporate guilds vanished in the face of the division of labour in each single workshop.'[16] Marx himself describes in the first volume of *Capital* what he sees as the two origins of the factory system of manufacture. The first was the bringing together of a number of craftsmen with specialized skills, each traditionally carrying out a different part of the same production process, under the control of a single capitalist; the two examples he gives of this are coach-building and the manufacture of cloth. Alternatively, Marx describes the situation in which a number of craftsmen, each carrying out the same task, were employed in a workshop under a single unit of capital. Originally each craftsman followed the old methods of handicraft, but 'before long, external conditions render it necessary that the concentration of workers in one and the same spot, and the simultaneity of their labour, shall be utilized in a different way'.[17]

As Marx points out, it was in the early workshops that the origins of the industrial division of labour are to be found. Those who ran the workshops soon realized that rather than having a number of people each perform simultaneously a set of relatively complex tasks, far greater efficiency could be achieved by splitting the production process into a number of simpler consecutive tasks. Each individual was then assigned one particular task, and the article or commodity being worked on was passed from one individual to another along the 'production line' until it had been completed. Adam Smith in his *Wealth of Nations*, published in 1776, cites a famous example of the advantages to be derived from the division of labour in the manufacture of pins. He points out that ten pin-makers, by dividing the job into separate tasks, are able to produce 48,000 pins in the space of time that, working individually, they might only have been able to produce ten. Smith writes that 'the greatest improvement of the productive powers of labour, and the greater part in the skill, dexterity and judgement with which it is directed, or applied, seem to have been the effects of the division of labour'.[18] He adds that it was this that brought about, in a well-governed society, 'that universal opulence which extends itself to the lowest ranks of the people'.[19]

The division of labour and the general fragmentation of production tasks, both within a single firm, and between different firms in the same field, is still the prevailing pattern of contemporary industrial production. The techniques have been considerably refined since the days of Adam Smith. In Britain, the optimal work cycle is now considered to be one minute, and in the US there have been efforts to bring it down to fifteen seconds. The basic organization of any production-line is still largely determined by what management have calculated to be the most 'efficient' ways of fragmenting tasks. Contemporary production technology reflects this situation. According to Galbraith, the most important consequence of technology, 'at least for purposes of economics, is in forcing the division and subdivision of any (practical) task into its component parts. Thus, and only thus, can organized knowledge be brought to bear on performance . . . nearly all of the consequences of modern technology, and much of the shape of modern industry, derive from this need to divide and subdivide tasks'.[20]

However we are jumping ahead a bit. First we must return to a second important step in the history of European technology, the introduction of the machine into factory manufacture. The largest source of power in the Middle Ages was the water-mill. First developed by the Greeks to grind corn, it was used for additional purposes, such as sawing wood, as early as the fourth century AD. By the fourteenth century it was being used for manufacturing in almost all the major industrial centres of Europe. There are accounts of water-mills providing power for pulping rags for paper in 1290, and for hammering and cutting the machines of an ironworks in 1320, as well as being used for purposes such as beating hides in a tannery, providing power for spinning silks, working up the felts in fulling-mills, and turning the grinding machines of armourers.

The replacement of human energy by mechanical energy in production was an important step. Not only did it greatly increase productivity, but the separation of energy from skill immediately gave an impersonality to the whole production process. Both the machine-tool and the machine developed according to the possibilities offered by the new engines of power. A number of important mechanical inventions developed in the early stages of the Industrial Revolution used the power made available by water-mills. Richard Arkwright's spinning frame, for example, produced in 1768, required more power than could be provided by a human operator, and from the beginning the work was carried out in mills. Water-power, however, presented a number of disadvantages. In the winter the streams froze, and the mills ground to a halt; similarly in times of drought they were unable to operate. Because there was no means of transmitting the power which they produced over large distances, the factories had to be placed near fast-running streams and rivers, which partially explains the growth of early textile industries in relatively hilly areas such as Lancashire and Yorkshire. This presented many problems of transport and general communication; raw materials often had to be brought to the factory from a considerable distance, while the finished products similarly had often to be taken to markets many miles away.

The major innovation that altered this situation was the steam engine. For the first time, it was possible to obtain the large quantities of energy required to power machines without having to rely on natural forces. Factories need no longer be situated near fast-running streams (although many retained the name 'mill'). The geographical mobility provided by the steam engine made possible the rapid growth of industrial towns all over the North of England. Again, however, it would be wrong to put this innovation down to chance. Ashton, who has stressed that the foundations of the Industrial Revolution were as much economic as they were technical, has pointed out the other activities and interests that were necessary before steam-power could be introduced. 'The researches of Blake, the capital and enterprise of Boulton, the ingenuity of Wilkinson, the technical skill of Southern and a host of obscure artificers were all necessary to the making of the steam engine', he writes.[21] Watt's experiments were in fact financed by one of the first major industrial capitalists, John Roebuck, and the two shared the first patent taken out by Watt in 1769. According to Ashton, 'the conjuncture of growing supplies of land, labour and capital made possible the expansion of industry; coal and steam provided the fuel and power for large-scale manufacture; low rates of interest, rising prices, and high expectations of profit offered the incentive'.[22] The Industrial Revolution resulted from the interaction between these three fields.

Marx and Engels describe the same process in the *Communist Manifesto*. As markets for industrial products continued to grow, they point out, elementary manufacturing techniques were unable to keep up with the growing demand. 'Thereupon, steam and machinery revolutionized industrial production. The place of manufacture was taken by the giant, modern industry, the place of the industrial middle class by industrial millionaires, the leaders of whole industrial armies, the modern bourgeois . . . the market has given an immense development to commerce, to navigation, to communication by land. This development has, in its turn, reacted on the extension of industry.'[23]

Most economic and social historians would agree with the implied characterization of the historical development of technology. They would find little to argue with in the interpretation of the growth of industrial society resulting from the introduction of machine manufacture. In other words, the growing demands of the market are seen as providing the primary incentive for technological innovation through the need for more efficient and productive machinery. At the same time, the nature of class divisions at any particular historical instance – referred to by Marx as the 'relations of production' – is itself held to be a reflection of the level of technological development. Medieval agricultural and production techniques required a feudalistic society – or so this interpretation goes – in order to be applied effectively. The emergence of factory production required the concentration of capital in a few hands, and simultaneously provided the means by which this capital could be increased most effectively. And so on.

Social development was seen as the natural progression through a number of states, from nomadism and primitive communism through feudalism to industrial capitalism. Technology provided the tools by which this development was to be achieved, but played a passive and neutral role in the development process itself, which was left to the operation of economic forces. Machinery was for Marx primarily a means of creating surplus value, the term he used to describe the excess of the market value of the final product over the cost of raw material used in manufacture and the wages paid to cover any human labour involved. 'The working machine is a mechanism which, through the instrumentality of the tools attached to it, carries out the very same operation which the manual worker of former days carried out with tools of a like kind',[24] writes Marx in *Capital*. 'If we regard machinery exclusively as a means for cheapening the product, the limit to its use is that its own product shall cost less than the labour which is replaced by it . . . the use of machinery is limited by the difference between the value of the machine and the value of the labour power replaced by it.'[25] (It is important to

53

emphasize, however, that although Marx realized the crucial importance of the development of material forces of production under capitalism, this is far from making him a technological determinist. He repeatedly stresses that it is not technology which makes it necessary for the capitalist to accumulate, but the need for accumulation which makes him develop the powers of technology.)

The early years of the nineteenth century saw the beginnings of what we have labelled the 'ideology of industrialization'. The social squalor of the late Industrial Revolution was still to come, and industry still had an air of transcendental power about it. Artists such as Joseph Wright of Derby and Philip de Loutherberg painted the factories with almost mystic reverence. Turner based his euphoric 'Rain, Steam, Speed' on observations made in torrential rain leaning out of the window of a train crossing Maidenhead Bridge. The same fervour for industrial development gripped the early capitalists. An outbreak of 'railway mania' led to a boom in railway shares in the 1840s – and a subsequent crash in 1849. The Great Exhibition of 1851, perhaps the pinnacle of the Industrial Revolution, became almost a religious festival in celebration of the powers of industrialization.

The idea that the processes of industrial production develop according to an internal logic of their own, providing merely the tools for economic development independent of political motive, was later taken up by Lenin. In his *State and Revolution*, written one month before the Bolshevik seizure of power in October 1917, he equated modernization with the set of industrial techniques already being used by capitalist countries. The creation of what appeared to be general material abundance resulting from industrialization was a theme subsequently embraced by the Bolsheviks. Since such industrial technology was not available to any great extent in Russia before 1917, Lenin decided that the forthcoming 'cultural revolution' would require full-scale industrialization and expansion of machine technology on the capitalist model. Communism was to be achieved through the integration of socialism and electricity, and this meant literally that the whole of Russia should be equipped with electric power. 'We ourselves, the workers,' he wrote, 'will organize large-scale production on the basis of what capitalism has already created, relying on our own experience as workers, establishing strict, iron discipline supported by the state power of the armed workers . . . Such a beginning, on the basis of large-scale production, will of itself lead to the gradual withering-away of all bureaucracy.'[26] Technical experts were still to be used, along with foremen and book-keepers; the main difference was that they were to be employed by the workers rather than by the capitalists.

The general attitude in Russia towards industrialization at this time is illustrated by the many Bolshevik revolutionary posters which, like many British paintings of the Industrial Revolution, betray an almost mystical belief in mechanical power. Large factories belch forth huge clouds of black smoke, a symbol of modernization and industrial strength not only in Russia, but also in Japan, Europe and North America. In art, these euphoric attitudes were reflected in the activities of the Russian constructivists, whose leading figures, such as Tatlin and Malevich, were intent on symbolizing the new machine age that was being greeted with so much fervour. 'This is our century, machine-technology-socialism. Come to terms with it, and shoulder the tasks of the century', wrote the Hungarian artist Moholy-Nagy a few years later.[27]

Lenin himself was particularly enamoured of the work-study techniques that had been developed in the 1880s by an American engineer, Frederick Winslow Taylor, under the label 'scientific management'. Taylor's basic approach was to apply to manual operations the principles that machine designers had learned to apply to the work of a tool during the early part of the nineteenth century. He broke a task down into its component parts, and then rearranged and re-integrated those parts in the most 'efficient' way. This was a major advance on the more straightforward division of labour; it indicated to management not only what each worker should do, but how he should do it.

Taylor first developed his ideas after studying the activities of labourers loading pig-iron for the Bethlem Steel Company in the US. By issuing precise instructions to the workers on how they were expected to carry out their work, based on his own observations and calculations, he managed to increase the daily load of coal per man from 16 tons to 59 tons a day, and the loading of bricks from 120 to 350 per man per hour. The Bethlem Steel works was able to reduce its work-force from five hundred to one hundred and forty. Although wages were increased by 60 per cent – financial incentives were an important way of obtaining high rates of working by those who were lucky enough to keep their jobs – the company received a considerable financial benefit. Taylor's ideas launched a fashion for 'industrial' engineering, particularly in the United Kingdom and the US, between 1910 and 1930, resulting in techniques that have since become an essential part of industrial processes almost everywhere. In particular, what we now know as 'automation' is conceptually a logical extension of Taylor's scientific management.

To Lenin scientific management appeared to be not only a useful tool for capitalism, but the answer for socialist production too. He seemed to ignore the fact that Taylor's whole approach involved treating men as if they were no more than unthinking and unfeeling machines; a necessary condition for efficient implementation of his ideas, Taylor wrote, was that the worker 'shall be so stupid and so phlegmatic that he more nearly resembles in his mental make-up the ox than any other type'.[28] According to Lenin in 1918: 'The Taylor system, is a combination of the refined brutality of bourgeois exploitation and a number of the greatest scientific achievements in the field of analysing the mechanical motions of work . . . The Soviet Union must at all costs adapt all that is valuable in the achievements of science and technology in this field . . . We must organize in Russia the study and teaching of the Taylor system and systematically try it out and adopt it to our own ends.'[29]

In a previous article, written in 1914, he had condemned the way in which Taylorism was used under capitalism, but had refrained from attacking the techniques themselves which made possible what he called 'the rational distribution of social labour'.[30]

A general belief in the functional interdependence model of technological development, and hence in the ideological neutrality of the technology itself, goes far towards explaining the duplication in Russia and the socialist countries of Eastern Europe of the patterns of industrial development followed in the Western capitalist countries. There are also many ways in which it remains manifest in our own society. One example is the general equating of economic growth with social development, and the idea that the former can be most efficiently achieved through massive invest-ment in the most advanced and highly industrialized sectors of the economy. As mentioned in the previous chapter, this is one of the main concepts behind the development strategies promoted by many development economists, now being applied to the problems of Third World countries.

Even where an emphasis on the importance of aggregate economic growth is beginning to falter, it does not necessarily signify any general change in attitude towards the possibilities offered by technology, only an indication that we may have to make our techniques of application more sophisticated. The Organization for Economic Co-operation and Development, for example, published a report in 1972 entitled *Science, Growth and Society*. This report highlighted some of the reasons for diverting research and development effort away from a concentration solely on means of achieving economic growth. In his introduction to the report, Professor Harvey Brooks, Dean of the Division of Engineer-ing and Applied Physics at Harvard and chairman of the com-mittee which prepared the report, outlined some of the basic assumptions that had been made. 'In general, they represent an extrapolation away from defence, national prestige and quanti-tative economic growth towards the more rational management of both growth and new technology in the interests of newly per-ceived social needs and increased effectiveness in the performance of the public sector and the fulfilment of collective needs.'[31]

In other words, the problems now associated with advanced technology are not interpreted as meaning that we have been wrong to try to use technology to solve all our social problems. It is just that we have not been sophisticated enough in our approach, that is we have not always selected a technology 'truly' appropriate to a particular social and economic situation. The basic idea of social development guaranteed by cumulative technological change remains intact. It even appears to take on an added importance, as we now need the services of a 'social technology' to predict what type of innovations we should make in productive technology. Hence the growing interest in techniques of tech-nology assessment, by which attempts are made to assess the social consequences of a particular technological innovation so that the less desirable aspects of any particular innovation can be minimized. In his book *Future Shock*, Alvin Toffler provides an illustration of the type of faith that is now widely attached to technology assessment: 'A sensitive system of indicators geared to measuring the achievement of social and cultural goals, and integrated with economic indicators, is part of the technical equipment that any society needs before it can successfully reach the next stage of eco-technological development. It is an absolute pre-condition for post-technocratic planning and change-management.'[32]

If the problems created by advanced technology, as I shall attempt to show, are in part due to the same ideological assumptions that lie behind the assessment techniques, in particular the supposed 'objectivity' of the scientific method, then it does not take much intelligence to see that these techniques are unlikely to get very far. The Roskill Commission, set up in 1969 to examine the siting of London's third airport, received massive volumes of 'expert evidence'. This illustrated, for example, that it is less socially-damaging to fly loud aircraft over lower-class as opposed to middle-class areas because of the different effect on property values in the two areas. It is one of the more sophisticated examples of the 'techno-economic' approach to social engineering. Others can be found in any attempt at 'scientific' planning, where a combination of cost-benefit and systems analysis is brought in to apply the gloss of scientific objectivity – and hence economic rationality – to what are frequently the manifestations of political or ideological prejudice.

The general techniques of planning illustrate a further extension of the ideology of industrialization. Within the framework of industrial production, Taylor's ideas have been superseded by concepts such as 'job enrichment' produced by industrial socio-logists, concerned with the fact that, even though it is necessary to continue to treat men as machines in order to get them to work as efficiently as possible, we cannot ignore apparent physiological and psychological requirements of the worker, and indeed may be able to use these to increase the efficiency of production. The techniques of operational analysis and 'human engineering' have carried the sophistication of this process even farther. 'Man is a complex set of interacting mechanisms with extensive feed-back loops within and between systems', we now learn from the in-dustrial psychologists. 'He must also be considered as an on-going system continuously adjusting his pattern of internal and external activities so that the study of cause/effect relationships is essentially fallacious.'[33] Such characterizations of the individual in terms of a set of essentially passive and objective qualities provide adequate legitimation for techniques aimed at maximum exploitation.

The ideology of industrialization is also present in a wide range of attitudes to the technological needs of countries of the Third World. This issue will be dealt with a greater detail in a subsequent chapter. However, many examples of this approach can be quoted. One is the statement made by Robert Macnamara, President of the World Bank, that the cost of bringing up a child in the Third World is $600, but the cost of avoiding its birth through birth control techniques merely $6.

The fact that such a calculation may, within its own limited terms, be correct is beside the point. What is important is the general attitude it reflects towards the problems of development as seen through Western eyes, and the general framework within which they are conceived. Development is seen as a process which can and should be 'objectively' quantified in economic terms, even if, as the case with intermediate technology, a relatively sophisticated economic approach is used in an attempt to evaluate social needs. As Samir Amin, director of the UN's African Institute for Economic Development and Planning in Dakar, has written on the literature covering the relations between population growth and development, 'the increasing use of refined techniques to "measure" the relations, quantified in terms of "cost-benefit", for example, just as the magic use of computers, does not automatically make it scientific in nature. On the contrary, the dominant theme of this literature is based on ideology and not on science; and this ideology corresponds to the requirements of a political plan of action which hinders the development of the "Third World", especially that of tropical Africa.'[34]

This is the crux of the matter. The whole approach is in fact an ideology, a set of ideas which distorts historical reality and acts as a powerful means of disguising its political roots. Industrialization provides an apparent rationality for – and hence appears to legitimate – policies of an exploitative nature. It preaches emancipation through the machine, and has indeed been successful in raising the standard of living for many; yet at the same time it is used to justify the increased domination and oppression of man that has been made possible by the machine. It preaches the social equality and democratization that the machine will bring; yet although it has broken down many traditional class barriers, it is used to legitimate and promote new class divisions and in-equalities. Above all, industrialization preaches the political neutrality of technology, representing it as merely a tool to be used for good or ill; yet it produces a technology which is a direct reflection of the ideology of advanced technocratic society, namely the dominance of 'scientifically rational thought' and of authoritarian forms of social control over all other interpretations of human experience. According to Jürgen Habermas, 'What is singular about the "rationality" of science and technology is that it characterizes the growing potential of self-surpassing pro-ductive forces which continually threaten the institutional frame-work *and at the same time* set the standard of legitimation for the production relations that restrict this potential.'[35]

The ideology of industrialization has been used by both Right and Left in the traditional political spectrum. In a pamphlet entitled *Government and High Technology*, for example, Professor John Jewkes suggests that the British Government has proved itself incapable of the management of complex high technology, that political attempts to undermine the neutrality and inner logic of technological development must necessarily lead to unwarranted interference and attempts to plan the unplannable. He concludes that the Government should restrict itself to the support of pure research in the universities and the development of technology required by the public services, leaving the rest to the interplay of market forces.[36]

In a slightly different way, Engels used the apparent neutrality of technology to attack the anti-authoritarian attitudes of the Anarchists.

'Take a factory, a railway, a ship on the high seas – is it not clear that not one of these complex technical establishments, based on the employment of machinery and the planned co-operation of many people, could function without a certain amount of subordination and, consequently, without a certain amount of authority and power? If the autonomists confined themselves to saying that the social organization of the future would restrict authority solely to the limits within which the conditions of production render it inevitable, we could understand each other; but *they are blind to all the facts that make the thing necessary* and they passionately fight the word.'[37]

Even today, the processes of technological and political revolution are seen by most socialist countries (and official communist parties) to be autonomous and to work at different levels. The first refers to the development of the forces of production, the second to the relations of production; contradiction is said to arise from the conflicting potentialities of the two. To quote Dr Mikhail Drobyshev of the Moscow Academy of Science, speaking at a recent Unesco conference on the tensions between science and culture: 'The progress of the present-day scientific and technological revolution is being held back by capitalist production relations and certain cultural elements based on them. To take a simple analogy, one might say that the train of scientific and technical progress has already almost gone through the station of capitalism, and the passengers who have not got on it will hardly be able to jump on the footboard, however hard they run along the platform.'[38] Similarly the French Communist Party adopted a resolution in 1966 that 'The century in which we live . . . is that of mathematics, of physics, of chemistry and of biology, the century of nuclear energy and of space flights, of television and of cybernetics . . . the same system which exploits the workers . . . limits the emancipatory powers of science.'[39]

Industrialization, as all ideologies, is dressed up as an absolute and objective set of values that refuses to recognize the validity of any criticism. In many ways, it can be compared to primitive magic: maybe the techniques didn't always work, but this was probably due to a human error in attempting to communicate correctly with the supernatural, a mistake that could be amended next time round. In his book *The Savage Mind*, the French anthropologist Claude Lévi-Strauss characterizes the difference between magic and contemporary science as follows: 'Magic postulates a complete and all-embracing determinism. Science on the other hand, is based on a distinction between levels: only some of these admit forms of determinism; on others the same forms of determinism are held not to apply. One can go further and think of the rigorous precision of magical thoughts and ritual process as an expression of the unconscious apprehension of the *truth of determinism*, the mode in which scientific phenomena exist.'[40] Substitute the ideology of industrialization for magic, and we have a clue to some of our present problems. Particularly when we remember the words of another anthropologist, E. E. Evans-Pritchard, in describing the all-embracing witchcraft beliefs of the Azande tribe. 'In this web of belief every strand depends upon every other strand, and a Zande cannot get out of its meshes because it is the only world he knows. The web is not an external structure in which he is enclosed. It is the texture of his thought and he cannot think that his thought is wrong.'[41] As Robin Horton has commented on this and similar observations, 'absence of any awareness of alternatives makes for an absolute acceptance of the established theoretical tenets, and removes any possibility of questioning them'.[42] Again the parallel with ideological attitudes to industrialization needs little elaboration.

3 The Politics of Technical Change

We must now take a closer look at the process of technological innovation. In the previous chapter we have seen how what has been described as the ideology of industrialization provides us with a particular view of this process. This interpretation, I suggest, obscures the way in which technological innovation is intimately related to the issues of power and control in society by explaining it in the apparently neutral terms of increased efficiency and productivity. To reveal the political and social forces at work, it is not sufficient to look merely at the consequences of a particular technological development, or even at the economic conditions that made its introduction possible. We must go beyond these to examine the specific reasons leading to the process of innovation in particular instances. These were often mediated as much by the prevailing attitudes towards the nature of both man and society as to supposedly objective social conditions. We shall see how the nature of technology comes to reflect and legitimate the dominant modes of instrumental and productive activity of a society. And I hope to indicate how, as the economic historian David Landes has written on the effects of technological change during the Industrial Revolution, 'The reorganization of work entailed the reorganization of labour: the relationships of the men to one another and to their employees were implicit in the mode of production: technology and social patterns reinforced each other.'[1]

It would be a major undertaking to trace this idea and its implications throughout the whole of human history, and well beyond the scope of this short book. In this chapter I shall merely try to give an indication of how this process worked out in practice during a number of crucial periods in the history of Western technology. Each stage that I have taken represented an important step towards our present situation; and each shows how the dominant forms of social organization and control, under the mask of an appropriate ideology, became built into the technology of the time. The cumulative effect of these attitudes now confronts us in contemporary technology. In particular the means of production of industrialized societies since the Industrial Revolution have to a great extent become a reflection of the relations of production under which they were primarily carried out, the authoritarian and hierarchical class relations of industrial capitalism.

Our analysis will suggest how, under these relations of production, technological innovation was determined, not only by concern for the efficiency of production technology, but also by the requirements of a technology that maintained authoritarian forms of discipline, hierarchical regimentation and fragmentation of the labour force. Furthermore, technological consumer products – such as cars and television sets – have been designed within the framework, not so much of social desirability, but of the need for a capitalist economy to develop the means to sustain itself. Equally, contemporary technology has been developed on the basis of maximum exploitation of the natural environment, with little regard to the need to maintain a stable equilibrium. The history of technology can be seen to trace the historical transformation from the domination of man by nature, to the domination of man and nature by man.

The historical roots of technology lie in the primitive tools of tribal societies. Although such tools often displayed a mechanical simplicity, the way in which they were used could be highly complex. Whether we study tribes in New Guinea, the traditions of the Red Indians, or the activities of the Eskimo, we almost invariably find two important factors. The first is that the set of magico-religious values associated with different elements in the natural environment – species of animals, for example, or resources taken from the earth – often makes any interaction with this environment through the use of technology into a symbolically significant act. Digging the earth, gathering certain plants for food, or killing particular animals could each carry a significance far greater than the functional importance of the activity. In ancient Egypt, for example, any craftsman involved in metal-working or glass-blowing was thought to be interfering with the natural production of metals and ores. As well as paying attention to many taboos they had to propitiate the gods or spirits who guided their work with offerings and prayers. There is a Sumerian poem which praises the divine origin and virtues of the various tools of the craftsmen, and Assyrian texts instructing glass-makers, before lighting a furnace, to bury a fœtus below it.

The second point, often associated with the first, is that social production in tribal societies, and the religious rites associated with it, was almost invariably designed in a way that guaranteed the ecological integrity of the environment on which the society depended for its material resources, as part of the social integrity of the society itself. The Eskimos, for example, believed that all living animals had a soul. In religious rituals, the fragility of the Arctic ecosystem was reflected in the Eskimo's attempts to foster friendly relationships with the animals that he hunted for food and clothing; he was careful to avoid the hardship that might be invoked by offending the soul of the animal he killed.[2] The grazing habits of nomadic tribes in Afghanistan have been shown to be organized in a way that gives maximum benefit to the different groups succeeding each other on the same pasture. Dr Fredrik Barth, an anthropologist who has made extensive studies of such nomadic tribes, has written that 'the pattern of utilization of pastures is thus not one of wholesale invasion and evacuation of areas, but a continual process of movement by different groups, tending toward a rather subtle adaptation of rate of utilization to the rate of production of the pastures'.[3]

Similar ecological awareness was shown by the Plains Indians of North America, who relied for their main source of food on the vast herds of buffalo that used to roam the continent. They evolved a strict code of rules to ensure that only the minimum number of buffalo required for food was killed, leaving the herds intact from year to year. Even the shifting-cultivation patterns of crop agriculture practised, for example, by the Indian tribes of the Amazon basin had a sound ecological base. The practice involved the intensive use of a clearing made in a forest for two or three years until all the goodness had been taken from the soil. The system was based on the fact that the tribes then moved on to a new settlement, returning to their original site many years later, by which time the vegetation had been able to return to normal. In these and similar ways, we find that production techniques were – and occasionally still are – an integral element in the whole way of life of a tribal society. They can rarely be isolated from religious, social, cultural or ecological considerations.

Lévi-Strauss suggests that even the primitive technologies used by these societies must have required a certain type of abstract knowledge. Activities such as making water-tight pottery out of friable and unstable clay, or changing toxic roots and seeds into harmless foodstuffs, all 'required a genuinely scientific attitude, sustained and watchful interest and a desire for knowledge for its own sake'. He suggests that the main reasons why this particular type of science did not lead to rapid social and economic development was that it was fundamentally different from the type of science that we have today. He identifies two distinct modes of scientific thought, representing 'two strategic levels at which nature is accessible to scientific inquiry: one roughly adapted to that of perception and imagination; the other at a remove from it'.[4] The first he labels the 'science of the concrete', distinguishing it from the second, which is the abstract science we have today. (A similar distinction has been made by Joseph Needham between traditional Chinese science and contemporary Western science.)

An important aspect of this distinction lies at the operational level. Lévi-Strauss characterizes the primitive technologist as a *bricoleur* – a jack-of-all-trades or professional do-it-yourself man. Unlike the contemporary engineer, the *bricoleur* is not able to rely on raw materials and tools specially selected for a particular task. 'His universe of instruments is closed and the rules of the game are always to make do with "whatever is at hand", that is to say with a set of tools and materials which is always finite and is also heterogeneous.'[5] The engineer, according to Lévi-Strauss, is always trying to escape the constraints imposed by a particular state of civilization, while the *bricoleur* by inclination or necessity always remains within them. A similar point is made by Robin Horton in his study of the relationship between African traditional thought and Western science. The key difference, he writes, is a very simple one: 'It is that in traditional cultures there is no developed awareness of alternatives to the established body of theoretical tenets; whereas in scientifically-oriented cultures, such an awareness is highly developed. It is this difference we refer to when we say that traditional cultures are "closed" and scientifically-oriented cultures "open".[6]

In early tribal societies, technological innovation seems to have remained subservient to social norms themselves the direct product of collective experience. This is not to say that innovation did not exist, but that it was based on requirements themselves determined by social practice. Joseph Ki-Zerbo has described traditional black society as being 'in a perpetual state of invention. At a technical and economic level, each family, each village, each tribal group discovered the means for a positive equilibrium with nature. To see this, one only needs to look at the varieties of selected grain, cultural methods, the extremely varied tools and forms of work, the many cures developed, even if they were administered with abundance of magic ritual'.[7]

Technological developments therefore remained within the limits of the existing social and cultural traditions. Although developed from an immediate system of social labour and its stock of accumulated technically-exploitable knowledge, they were never allowed to extend to the point at which they might present a challenge to the prevailing pattern of social organization.

Both Lévi-Strauss and Horton emphasize that primitive technologies, based on a 'concrete' rather than an abstract science, and thus along supposedly 'closed' rather than 'open' lines, were usually developed according to implicit concepts of social and environmental stability. As an integral part of a general socialization process, they were developed in a way that related the individual both to other members of the tribal community, and to the surrounding environment. It is interesting to note here that industrial sociologists have discovered an equivalent phenomenon of opposition to technical change in 'stabilized' work groups in modern industry. In addition technological activity involving a number of participants is claimed to increase social cohesion between them when working as an identifiable group. An OECD report dealing with the attitude of workers towards technical change concluded that 'in a social environment in which the consciousness of the community is strong and occupation and social lives are closely interwoven . . . resistance to change in either sphere is strong. On the other hand, individuals who identify with growth and diversified societies are likely to have less reservations about innovation'.[8]

In tracing the history of contemporary technology and showing how it broke away from these early traditions, we need to start with the Middle Ages. It was during this period that a large number of technological developments made by earlier civilizations – not only the Greeks and the Romans but also the Chinese, Arabs, Indians and other cultures – first began to be welded together into the technological programme that later accompanied the emergence of industrial capitalism.

Many of the early developments in both science and technology began in the monasteries. The religious environment provided relative protection from the social turmoil of the outside world. It also provided an ethic that extolled the virtues of both abstract reflection and hard work. Monasteries often took the lead in mechanizing arduous tasks (possibly, it has been suggested, so as to leave as much time as possible free for devotional exercises). The Cistercian monks, for example, built many of their monasteries near rivers to enable them to set up water-mills; one such mill at Clairvaux Abbey in France made use of water power to grind corn, heat water, operate fulling machines, and carry away refuse, all this being done by the same water in successive steps. Similar uses of water-power can be found in Fountains Abbey in Yorkshire. As Mumford says, 'the spiritual routine of the monastery, if it did not positively favour the machine, at least nullified many of the influences that worked against it. And unlike the similar discipline of the Bhuddists, that of Western monks gave rise to more fertile and complex kinds of machinery than prayer wheels.'[9] Similarly, Whitehead has suggested that 'the alliance of science and technology, by which learning is kept in contact with irreducible and stubborn facts, owes much to the practical bent of the early Benedictines'.[10]

One immediate consequence of the rigid discipline of the monastery was the development of the clock, considered by many as the key invention of the modern industrial age. Time was an important part of monastery discipline. A bull of Pope Sabianus, for example, decreed that the bells of all monasteries should be rung seven times in the twenty-four hours of the day. It soon be-

came necessary to develop a means of keeping count of such can-
onical hours and ensuring their regular repetition. The first clocks
used in monasteries were only water clocks, but mechanical
clocks are known to have existed in the thirteenth century. Soon
bell-towers were springing up everywhere, announcing the regular
hours of the day; their prime function was to ensure and formalize
devotional obedience, but they also brought regularity into the life
of the workman and the merchant.[11]

Lynn White has recorded that 'suddenly towards the middle of
the fourteenth century, the mechanical clock seized the imagination of our ancestors. Something in the civic pride which earlier
had expended itself in cathedral building now was diverted to the
construction of astronomical clocks of astounding intricacy and
elaboration'.[12] There is a vital point here. The new production
technology of the machine emerged from the *double* discipline of
the monastery, the discipline of both mental and physical activity.
The former both legitimated and, through the development and
use of the clock, largely determined the latter. The science that
was developed in the monasteries was an abstract science based
on the philosophical traditions of the Greeks. It was very different
in concept from the 'science of the concrete' that had gone before.
It naturally gave rise to a very different technology, one which
translated the anthropocentrism of the Judaeo/Christian tradition
into machines that began to assert man's supposed dominance
over nature, and the application of natural power to his own ends.
The intellectual order contained in the new science becomes
synonymous, through the medium of religious belief, with the
social control that now came to be imposed on man's activities.
Such threads of thought later emerged in Methodism and other
religious movements that appeared to assert the divine purpose
of the Industrial Revolution. They were also to be found in the
subordination of human activity to the constraints of the clock
first developed by factory owners such as Josiah Wedgwood, and
later forming the basis of Taylor's 'scientific management'. Not
without reason can the Benedictines be identified among the
ancestors of modern capitalism.

Having already established a virtual monopoly of scholastic
knowledge, the authorities of the church attempted to extend this to
control over technical skills too. Cathedral building, for example,

absorbed a large proportion of the skilled work-force, and artists and artisans furnished internal decoration to match. A number of architects and sculptors who worked in a religious environment even took their working procedures – such as the principle of 'clarification' requiring the preliminary explicit arrangement of ideas in a general plan – directly from the traditions of Scholasticism developed in the monasteries. Many of the craft guilds were based on strong religious traditions and hence came under the powerful control of the religious authorities. Forty-seven of the 106 stained-glass windows in Chartres Cathedral, the earliest dating from 1194, were given by guilds, and many depicted the occupations of their donors.

Perhaps equally importantly, the church often set out in connection with the established professions, to destroy any technological activity that seemed to challenge its hegemony. Monks in England conducted an active campaign against hand-mills from the end of the thirteenth century until the Peasants' Revolt in 1381.[13] Women who attended childbirth were frequently the outcasts of society who could get no other work; with their herbal and medical knowledge often covering contraception and abortion, they later came to be accused of witchcraft. According to Margaret Murray, in her book *Witchcraft in Western Europe*, during the seventeenth and eighteenth centuries 'the Christian Church was still engaged in crushing out the remains of paganism and was reinforced in this by the medical profession, who recognized in the witches their most dangerous rivals in the economic field'. Similar opposition from the religious and medical authorities was experienced by one of the founders of modern chemistry, Paracelsus. He tried to argue that the chief purpose of the alchemist should be to help the apothecary and the physician, rather than devoting all his efforts to attempts at making gold. In 1525, five years after Luther had burnt the papal bull at Wittenburg in protest at the oppressive authority of the church of Rome, Paracelsus publicly burnt the works of Galen and Avicenna, the early alchemists, before the physicians of Basel, making a similar gesture in the name of science.[14] The strength of the opposition from the religious community to the ideas of scientists such as Copernicus, Galileo and Bruno – the latter being burnt at the stake for his religious scepticism – suggests the extent of the desire of religious authorities to retain a monopoly control over all areas of useful knowledge.

It was in the world of commerce that the machines developed under the stern discipline of the monastery were first applied on a broad social scale. The most crucial period from this point of view was during the seventeenth and early eighteenth centuries. It was predominantly in this period that groups of craftsmen were brought together into workshops, the beginning of factory production that became known as manufacture. This situation reflected the apparent economic need to organize production on a more 'rational' basis than had been done previously, in response to the increased opportunities for trade that were opening up.

The early workshops had two important implications. The first was that they signified the first emergence of the industrial capitalism. As Marx writes in *Capital*, 'A greater number of workers working together in one place . . . in order to produce the same type of commodity under the mastership of one capitalist, constitutes, both historically and logically the starting point of capitalist production.'[15] Equally important from our point of view were the mechanisms for maintaining discipline in the work situation that these early factories presented. And it was also during this period that the techniques of productive activity first began to take on their 'legitimating' role.

To understand more clearly the general sequence of events, we shall examine in some detail the case of the textile and weaving industries. These provide an example of the types of historical processes that were at work in all branches of productive industry. In addition the developments in the textile industries lay at the heart of the whole Industrial Revolution. Other examples such as the paper-making or pottery industry may differ slightly in detail, but the same broad principles behind technological innovation usually can be shown to apply.

Traditional weaving was a lengthy business. The wool or cotton had to be spun into a single, strong thread – or yarn – in which the fibres were lined up and twisted together. The traditional way of doing this was the spindle-and-distaff technique, which remained virtually unaltered for many thousands of years until the invention of the spinning-wheel, first introduced into Europe in the thirteenth century. Once the yarn had been prepared, washed and dyed, it was taken to the loom, on which it was passed through vertical threads already fixed in place – known as the warp – to make the weft. The basic design of the loom remained almost unchanged from the time of the Egyptians up to the Middle Ages. Various improvements were made possible by the increased power provided by water-mills and windmills from the eleventh century. Textile craftsmanship flourished in Flanders, Tuscany and parts of England between the twelfth and the sixteenth centuries. Each part of the overall process – spinning, carding, twisting, fulling – had its own craft guild. In general, however, these guilds were not closely linked to one another. Overall control of the whole production process usually lay in the hands of the bankers, merchants and financiers. These both supplied the necessary capital and raw materials, and also collected the yarn after it had been spun, delivered it to the weavers, and so on.

The financiers soon realized that, rather than going to the pre-dominantly town-based guilds to get the different processes carried out, there were a number of advantages to be gained from making use of the large pool of cheap labour that was available in rural areas. Here they were free from the restrictions imposed by the guilds on various aspects of production such as the exact nature of the final product, and the techniques which were to be used. 'Putting-out', as the new system began to be known, and the use of rural labour based on a system of cottage industry rapidly gained wide acceptance.

Demands for textiles escalated rapidly with the increased standards of living associated with growing patterns of trade and developments in other fields such as food production. The merchants, keen to exploit this market to the full, faced problems of maintaining a supply of goods. The life-style of the rural workers was an erratic one. On fine days they were as likely to be found ploughing the fields as working at their looms. Many trades honoured 'Saint Monday' as an additional holiday to Sunday, particularly in areas where small-scale domestic and outwork industries existed. The general pattern of work involved alternate bouts of labour and idleness. Engels, perhaps slightly biased, claimed that 'the workers vegetated throughout a passably comfortable existence, leading a righteous and peaceful life in all piety and probity'.[16]

The distaste of the merchant class towards such apparent indiscipline is widely documented. Burke's message to the labouring poor in 1795 was a simple one: 'Patience, labour, sobriety, frugality and religion, should be recommended to them; all the rest is downright fraud.' Such attitudes reflected an immediate problem faced by the merchants. Major advances in spinning techniques – the most important being Hargreaves' spinning jenny in the 1760s – had led to substantial increases in the amount of yarn that became available. The merchants had little control over the weavers, however, and thus their means of increasing productivity on a putting-out basis were limited. Practically their only means of control were financial ones, and frequent attempts were made to lower wages. But the weavers fought back. Often quantities of the merchants' materials were withheld, while weavers to whom looms had been hired realized that grievances against the merchants could be expressed directly by breaking them. In addition, the fact that the weavers, working in their own cottages, were often distributed over a large area of countryside led to many problems of transport and communication in general. It was frequently to get round these and similar problems that the merchants first began gathering the weavers into factories. Landes has described the predicament of the merchants in his study of technological innovation *The Unbound Prometheus*, quoted at the beginning of this chapter. 'One can understand why the thoughts of employers turned to workshops where the men would be brought together to labour under watchful overseers,' he writes, 'and to machines that would solve the shortage of manpower while curbing the insolence and dishonesty of the men.'[17]

It seems possible to identify four main reasons for the setting up of factories. The merchants wanted to control and market the total production of the weavers so as to minimize embezzlement, to maximize the input of work by forcing the weavers to work longer hours at greater speeds, to take control of all technical innovation so that it could be applied solely for capital accumulation, and generally to organize production so that the role of the capitalist became indispensable. Factories provided the organizational framework within which each of these could be achieved. Thus although machines were present in the early factories, they were seldom the *reason* for setting up a particular factory. The factory was a managerial rather than a technical necessity.[18] It imposed a new discipline on the whole production process, and was described by Charles Fourier as a 'mitigated form of convict prison'.

The weavers entered the factories with great reluctance. They resented the discipline that was being forced on them, having been used to organize their own hours of work. They also resented the way that the factory system affected family relationships. Previously spinning and weaving had been tasks that had involved the whole family, and had been carried out at home. Now the children were sent off to the spinning-mills, and the wives and adolescents to the power-loom sheds. According to the evidence of one witness before a Government Select Committee in 1834, 'All persons working on the power looms are working there by force, because they cannot exist any other way; they are generally people that have been distressed in their families and their affairs broken up . . . they are apt to go as little colonies to colonize these mills'.

But power machinery was expensive. Where the merchants could force the wages low enough – made easier once the power of guilds and workers' associations had been broken after a series of unsuccessful strikes – it still paid to use hand labour. E. P. Thompson has suggested that 'it can be argued that the very cheapness and superfluity of hand-loom weaving *retarded* mechanical invention and the application of capital in weaving'.[19]

The factory-based organization of the weaving industry therefore did not, as some historians imply, develop directly from a more efficient technical base; many of the new machines were developed and introduced only *after* the weavers had been concentrated into the factories. This is further underlined by the fact that other European countries, in which events such as the French Revolution had shown how concentrated labour could provide a powerful challenge to existing power structures, were much more reluctant to concentrate their weavers into factories. In these countries the putting-out system was retained, long after it had been almost abandoned in England.

A parallel example to the textile industry is provided by the efforts of Josiah Wedgwood in the field of pottery. It was Wedgwood who built up the pottery industry in the Midlands, establishing a large factory at Etruria near Stoke-on-Trent in the second half of the eighteenth century. Wedgwood was convinced from the very beginning that the only way he could obtain the standard and quantity of goods required by a rapidly expanding market was through a carefully calculated division of labour, involving the separation of all the different processes involved in pottery-making. This involved imposing a strong factory discipline. Wedgwood was one of the first industrialists to develop a clocking-in system. He also published a long and detailed set of instructions covering all aspects of factory discipline, as well as setting up the tradition of a foreman class.

Like the weavers, the potters were not used to this kind of discipline. 'The potters had enjoyed their independence too long to take kindly to the rules which Wedgwood tried to enforce – the punctuality, the constant attendance, the fixed hours, the scrupulous standards of care and cleanliness, the avoidance of waste, the ban on drinking,' as one historian has written.[20] Yet Wedgwood stuck to his principals. In doing so, he turned a collection of what he had called 'dilatory, drunken, worthless workmen' into what he described ten years later as 'a very good set of hands'. Etruria provided one of the first examples of the factory system which, as P. Mantoux says in his classic *The Industrial Revolution*, 'concentrates and multiplies the means of production so that the output is both accelerated and increased'.

The important fact about the beginning of the factory system – and hence of capitalist production in general – is that it does not seem to have been determined solely by technological or even economic factors. The Industrial Revolution emerged equally from the class relations of capitalism – the division of society into a capitalist and a proletarian class – that had begun to establish themselves with the rise of the merchant class well before any major advances in production technology. It is thus impossible to say that the relations of production were an inevitable consequence of improvements of the means of production; the one was a reflection of the other.

In particular, the political dimension embraced a new attitude towards the nature of man and his role in society, and the emergence of new ideas of the importance of work. According to Landes, the frustration and vexation of employers towards the labouring poor in the late seventeenth and early eighteenth centuries was reflected in the fact that 'where once poverty had been looked on as an unavoidable evil and the poor man as an object of pity, now poverty was a sin and the poor man a victim of his own iniquity'.[21]

The machines required in the early years of the Industrial Revolution were those which not only replaced hand-labour but also compelled the concentration of production into factories. A Dutch loom which could weave 24 narrow tapes simultaneously, and a complex hand-run knitting frame for weaving hosiery, both eminently suitable for domestic manufacture, were soon by-passed in favour of larger machines, whose mechanical superiority slowly eliminated the traditional forms of hand production. The main examples of these were Arkwright's water-frame, produced in 1768 and designed to make use of water-power for spinning cotton; Cartwright's power-loom, designed in 1784, which could be powered by horse, water-wheels or steam engines; and Crompton's mule – so called because it was a cross between Hargreave's spinning jenny and the water-frame – developed in 1779 and capable of producing a strong, fine and even yarn suitable for many types of textiles. The comparatively large output of these machines meant that they soon overstretched the capacity of the small streams that ran the mills. In 1785 the last logical step was taken by the adaptation of Watt's steam engine to providing power for such machines. Each of these developments was crucial to the setting up of the factory system, and contributed to the general disciplining of the work force. According to Ashton, 'it was only under the impact of powerful forces, attractive or repellent, that the English labourer or craftsman was transformed into a factory hand'.[22]

These were the forces that were legitimated by what we have called the ideology of industrialization – or, to put it another way, the 'price' to be paid for progress. They signified the emergence of the hierarchical and authoritarian power relationships of industrial capitalism. The success of the Industrial Revolution was as much due to changes in business management as it was to technological developments. Hartwell has commented that 'the character and success of those who survived the very high failure rates among the early industrialists can be explained, not so much in terms of differences in the quality of labour or of machines, but more in terms of differences in the quality of management'.[23] He also points out that one man often combined two or more of the roles of capitalist, inventor, innovator and manager. This reveals the close links that often bound these activities. Richard Arkwright – a fiercely strict employer and a technological plagiarizer – is perhaps one of the best examples of both a technological and managerial innovator.

Not everyone surrendered to the emerging forces of industrial capitalism. The end of the eighteenth century saw the growing activities of various groups of machine breakers, who set out to challenge the advances of the new technology. There were two different types of machine-breaking activity. To some, it became an established way of putting pressure on employers as part of a general pattern of class conflict. Typical of these disputes were the wrecking of coal-mines during widescale rioting in Northumberland in 1740, frame-breaking in the East Midlands hosiery trade, and the activities of Luddites – various groups of machine-breakers named after their legendary leader, Ned Ludd – in Nottinghamshire, Leicestershire and Derbyshire. Such activities had a dual purpose. They both confronted the employers directly, and stimulated the solidarity of the work-force. In the main, according to Eric Hobsbawm, machine-breaking frequently arose from the typical social relationships of capitalist production that were beginning to emerge as part of the Industrial Revolution. 'The men of 1760,' he writes, 'were still a good way from understanding the nature of the economic system they were about to face. Nevertheless it is clear that theirs was not a simple fight against technical progress as such.'[24]

Other machine-breaking activities contained a genuine hostility to the new machines, and the social relations of production that they brought with them. Machines threatened, not only employment, but a whole way of life that embraced the freedom, dignity and sense of kinship of the craft-worker. Attempts to challenge the machines met with wide support, not only from those who suffered directly, but from other classes too. The protection given by farmers threatened by the new agricultural machines partly explains why the Government found it necessary to deploy 12,000 troops to tackle the Luddite problem between

1811 and 1813, greater than Wellington's army in the Spanish Peninsula.

Opposition to the new machines in particular, rather than to mechanization in general, is illustrated by the activities of the Lancashire machine-wreckers of 1778 to 1780. They distinguished between spinning-jennies of 24 spindles or less, suitable for domestic production, which they spared, and the larger ones, suitable for use only in factories, which they destroyed. Similar ideas were later expressed by Samuel Butler in his novel *Erewhon*, first published in 1872. He adapts Darwin's ideas of natural evolution to the development of machines, suggesting that machines may eventually take over from man, and should therefore be destroyed. But he adds that: 'Man's very soul is due to the machines and their existence is quite as much a *sine qua non* for his, as his for theirs. This fact precludes us from proposing the complete annihilation of machinery, but surely it indicates that we should destroy as many of them as we can possibly dispense with, lest they should tyrannize over us even more completely.'[25] Although these words are put in the mouth of an Erewhonian philosopher, Butler later wrote in an introduction to the novel that he believed the theory 'to be quite sound'.

The machine-breakers won a number of local conflicts. Opposition to machinery in Norfolk, for example, was successful in keeping up the wages of weavers for a number of years. But the Luddites were powerless when faced with the gathering political momentum of industrial capitalism, supported both by strong military forces, and by the revolutionary élan that industry brought with it. 'By the time further technological advance once again changed the composition of the labour force', notes Landes, 'a new generation had grown up, inured to the discipline and precision of the mill.'[26]

Luddism faded away in the 1820s, giving way to more organized activities through trade unions and other early forms of workers' organizations in the factories. By imposing the discipline of the factory, the factory owners had unwittingly created a new political force, the industrial proletariat. The more that work-tasks were fragmented, the greater became the desire of workers to join together in united industrial action.

The next crucial stage in the development of technology came about at the same time, largely in response to this situation. Although labour militancy was usually directed at specific aims – such as better wages and an improvement in working conditions – it seriously challenged the efficient functioning of the vast industrial machine that was getting under way. Confrontation between the factory owners and the organized work-force became increasingly frequent. In this situation, technological innovation took on a new role. Machines began to be introduced not merely to help create a framework within which discipline could be imposed, but often as a conscious move on the part of employers to counter strikes and other forms of industrial militancy. Even the threat of mechanization, with the implied unemployment that this would bring, was often used by the employers to keep wages down. Thus although machinery for combing wool was available in 1825, the cheapness of the woolcombers' labour made it possible for the factory owners to keep the threat of machinery above their heads for a further twenty years.

'The need for industrial peace,' writes Thomson, 'for a stable work-force, and for a body of skilled and experienced workers, necessitated the modification of the (more brutal) managerial techniques – and indeed, the growth of new forms of paternalism – in the cotton-mills by the 1830s.'[27] Technological innovation was one of the new management techniques. Economic factors, although of primary importance in the long run, often became subordinate to short-term needs for dealing with labour. Profits could only be guaranteed and maintained with a relatively docile work-force. Machines frequently provided the means of achieving this docility. The introduction of machines became as much part of the day-to-day tactics of the class-struggle between labour and capital as it was part of the overall strategy. These tactics inevitably included the need for increased social control on the part of capital, and the authoritarian relationships that this implied became crystallized in the machines that were introduced.

To see this process at work, we need go little farther than a remarkable volume, *The Philosophy of Manufactures*, written by a Scottish academic, Andrew Ure, an arch-apologist for the whole factory system, and published in 1835. Ure describes vividly how manufacturers, oppressed by militant unions and 'unable to control workers by reducing wages', were led to use technological innovations for this purpose. A typical example was increasing the length of the spinning mules. By decreasing the overall number of mules required, this displaced adult spinners and increased the number of their assistants, thereby weakening the factory apprentice system and reducing the spinners' authority. 'This necessity of enlarging the spinning frame has recently given an extraordinary stimulus to mechanical science', writes Ure. 'In doubling the size of his mule, the owner is able to get rid of in-different or restive spinners, and to become once more master of his mill, which is no small advantage.'[28] This was despite the fact that such modifications were often costly, and that the larger machines meant that the lay-out of factories often had to be re-planned at considerable expense.

Another example quoted by Ure of the use of technological in-novation as a management tactic is the development of the self-acting mule. According to Ure, strikes in factories in various towns in the Midlands led local factory-owners to ask a firm of machinists in Manchester 'to direct the inventive talents of their master, Mr Roberts, to the construction of a self-acting mule, in order to emancipate the trade from galling slavery and impending ruin . . . Mr Roberts, who was then little versed in spinning machines, suspended his professional pursuits as an engineer and set his fertile genius to construct a spinning automaton'.[29] Richard Roberts was eventually successful in his efforts, and the self-acting mule was patented in 1830. £13,000, most of it supplied by the factory owners, had been spent on perfecting its design. Although slow at first to become accepted, factory owners soon found the new machine provided a way round the high wages demanded by the skilled spinners. Roberts became a rich man, while factory operatives got to know the dreaded new machine as 'the Iron Man'.

To Ure, the message was clear. 'This invention confirms the great doctrine already propounded, that when capital enlists science in her service, the refractory hand of labour will always be taught docility.'[30] It was this 'refractory hand' that had already destroyed John Kay's house in 1753, Hargreaves' spinning jennies in 1768, and Arkwright's mills in 1776 and subsequent years. It had also organized widescale spinners' strikes in 1818, during which shuttles had been locked in chapels and workshops in

Manchester, Barnley, Bolton, and other towns. The strikes were not necessarily confined to the textile industry, and each such strike provided an added incentive for mechanization. According to Samuel Smiles, the Victorian biographer of a number of eminent industrial figures, 'In the case of many of our most potent self-acting tools and machines, manufacturers could not be induced to adopt them until compelled to do so by strikes. This was the case of the self-acting mule, the wool-combing machine, the planing machine, the slotting machine, Nasmyth's steam-arm and many others.'[31]

History shows the success of industrialization during the Industrial Revolution in raising productivity and in the long run, wages and the general standard of living. But as we have seen, it would be wrong to interpret the contribution of machines to this solely in terms of increasing the efficiency of production. Their general contribution to 'social order' appears to have been equally important. Marx later confirmed this when he wrote that machinery 'is utilized as the most powerful weapon in the capitalist arsenal, as the best means for overcoming the revolts against capital'.[32]

A possible economic indication is provided by the concept of the 'economic residual'. Economic historians, on re-evaluating the process of economic growth during the Industrial Revolution, have discovered that the increase in the total value of the output of British industry over this period was considerably higher than the increase in value of the traditional inputs of land, labour and capital. One explanation is that the residual represents the contribution of technical skill and knowledge to greater productivity. Alternatively, it might be argued that it represents the greater economic efficiency afforded by the new machines in manipulating scarce resources. Part of this might be claimed to result from increasing patterns of authoritarian control over the labour input and its organization, the process referred to by Weber as 'rationalization'. In particular, this was achieved by an educational system designed to produce 'human capital' that would fit both the technical and the disciplinary framework of the emergent factory system.[33]

Machines might therefore be looked upon as the means by which the capitalist class reduced the entropy – or tendency to disorder – of the whole productive system. In this way it increased the effectiveness of the system through the imposition of an apparent technological 'order' or 'rationality' on the work-force. The resulting increase in organizational efficiency could result in raising the proportion of output to input, again appearing to legitimate the authoritarian and hierarchical relations of production that accompanied this process. Landes himself puts forward the tentative hypothesis that 'the factor-cost pattern required for a technological breakthrough is different from that needed for exploiting the possibilities of that breakthrough'.[34] In other words, we must look beyond mere economic factors, and hence concepts of functional interdependence between technological and economic development, in our historical interpretation of the process of technological innovation. A possible alternative, as I have already suggested, is to interpret this process as the reflection of the hierarchical and authoritarian class relations of industrial capitalist production.

In the way that science, as indicated in the first chapter, becomes the legitimating ideology of power, technology becomes the legitimating ideology of social control. Innovations in technology were never neutral, but were part of the political process itself. Marx writes in *Capital* that 'machinery possesses, as capital and through the instrumentality of the capitalists, both consciousness and a will; it is therefore animated with an urge to reduce to a minimum the resistance offered by the natural but elastic limitations of the human material through which it works'.[35] There is little point in introducing machines capable of increasing the efficiency of production without sufficient control over the workforce to ensure that the machines will be operated to their maximum capacity. Marx himself locates this control almost entirely in the hands of the overseer. 'In place of the slave-driver's lash,' he wrote, 'we have the overlooker's book of penalties.'[36] But the machines themselves also provided a source of control, as they reflected and supported the fragmentation and hierarchical organization of work tasks implied in the capitalist division of social labour.

This point has been also brought out by J. L. and Barbara Hammond, writing in the early years of this century on the social aspects of the Industrial Revolution. 'The new industry increased power to a remarkable degree,' they write in *The Town Labourer*, 'and it seemed to (the new) oligarchy the most natural thing in the world that the economic should resemble the political power, and that in the mill, as in the State, all this power should be concentrated in the hands of a few men, who were to act and think for the rest.'[37] In other words, the social relations between the groups of individuals employed in industry became the model by which the industrial state was to be run. Discipline and hierarchical control were the orders of the day. 'To all the evils from which the domestic worker had suffered,' write the Hammonds, 'the Industrial Revolution added discipline, and the discipline of a power driven by a competition that seemed as inhuman as the machines that thundered in factory and shed.'[38] Again we see the way the technology of factory production reflected the social relations of capitalist society, united in a common ideology. 'The animal machine – breakable in the best case, subject to a thousand sources of suffering – is chained fast to the iron machine, which knows no suffering and no weariness,' wrote James Philip Kay in 1832.

Industrial development in the United States began by following a pattern similar to that of Britain and Western Europe. Technological innovation and the growth of the factory system soon challenged existing craft traditions, although these often had a weaker hold than they did on communities in Europe. But it was the social revolution that accompanied these technological changes that primarily affected the industrial worker, and against this that his protests were made. Norman Ware writes: 'Aside from a few riots in Pittsburgh, in which the hand-loom weavers destroyed some of the new machines, the American worker was not opposed to the new machinery. He was opposed to the method of its introduction, for exploitative purposes, as he conceived it, in the hands of a group alien to the producer.'[39] For every protest against machine industry, Ware writes that a hundred can be found against the new power of capitalist production and its discipline.

In general the power of the early US unions was weak, and the relative lack of opposition to the new production technologies was an important factor in giving American industry a boost in its competition with Europe. Capital-scarcity, rather than labour-scarcity, seems to have been the main problem. Higher interest rates reduced the amount of capital available for each worker. Again, technological innovation was used as a lever to adjust the relationships between labour and capital. According to Nathan Rosenberg, 'It was not the high level of wages as such, but rather the persistent pressures on the labour market, the numerous opportunities for labour in a resource-abundant environment, and the high degree of labour mobility, that . . . gave [entrepreneurs] a strong bias towards the development of labour-saving techniques.'[40] In general an important function of management was to predict and adjust for any event that might obstruct the continuous flow of production. If labour was likely to present one such risk, then this could be countered, and if possible forestalled by adequate technological innovation.

Many of the tendencies that we have noticed during earlier periods of technological innovation are still present in the process today, even if manifest in a more sophisticated and complex way. Taylor's general approach to scientific management has been expanded to the 'rational planning' of almost all sections of the economy. The greatly increased importance of technology in modern society has led to a fusion of scientific research, the technological application of the results of research, its industrial utilization and its institutionalization in the category of research and development. One consequence is that the process of innovation has itself become institutionalized. No longer can industry rely on isolated inventors coming up with the right ideas at the right time. Competition and the need for continuous innovation have made it necessary for any major industrial company to run its own research and development division. This can consume large amounts of the company's annual turnover.

Technological innovation is still used by capital to maintain and strengthen its hold over labour. Containerization, for example, has been introduced into British ports partly to counteract the increasing militancy of the dockers and their success in pressing for higher wages by reducing the need for human labour. Part of the attraction of box-girder bridges is said to be the fact that, as in factory building, by having the pre-fabricated parts merely slotted together on site, the labour force is fragmented. This greatly reduces the chances of militant industrial action on building sites. According to a recent advertisement for a firm specializing in factory-based building methods, 'We've no problems of scattered supervision (of the labour force). So quality's always up to scratch.'

The threat of technological unemployment is also a means by which management can keep down wages. A number of employers have been accused of using the threat of replacing skilled female workers with automatic machines as a way of avoiding – and at the same time legitimating this avoidance – the legal requirement of putting female operators on equal rates of pay to male operators. The author of one OECD report mentions that 'technical procedures . . . remain the most important factor in determining the type of structure and the type of relations in a firm',[41] and also that 'change is not a pure fact. It is the manifestation of intentions, social relations, economic necessities and political projects'.[42] If part of the political project is to retain authoritarian and hierarchical control over the work-force, then this will be reflected, as we have seen, in the process of innovation, and the technology that emerges from this process.

It is impossible to isolate technological innovation, however, as the single factor which makes social control possible. Equally important are the whole range of management techniques that have been developed since Taylor's time, each associated with a particular approach to the production process. Taylor himself confessed to two aims. Not only did he desire to increase productivity, but he was also keen to reduce what he saw as indiscipline in the work force. The ideological contents of his approach is revealed by his avowed aim to 'render labour unions and strikes unnecessary', while admitting his belief that 'more docility produces higher wages'.

The importance of Taylorism is precisely the way in which a supposedly scientific approach is used to legitimate a particular set of management policies. Not surprisingly, Taylorism met with strong opposition from unions in the US at the turn of the century. The labour movement denounced the 'sweating and speed-up practice' that the 'cult' of scientific management encouraged because it 'looks upon the worker as a mere instrument of production and reduces him to a semi-automatic attachment to the machine or tool'.[43]

In fact, many of Taylor's own observations have been shown by later investigators to be wholly unscientific. In particular, the individual selected by Taylor to perform the tasks which he measured with a stop-watch for some of his most crucial observations was known for his large size, massive strength and general stupidity, factors that made him particularly suitable for the type of work pattern that Taylor was trying to establish. To extrapolate from the activities of this one individual to the organization of an entire work-force, *regardless of its effectiveness in increasing productivity*, cannot by any stretch of the imagination be called scientific. The prime function of the label seems to have been to give Taylor's ideas the stamp of phoney objectivity.

A similar unscientific attitude has been revealed by re-examination of studies that led to the emergence of ideas of 'job enrichment'. The most important of these were the so-called Hawthorne Studies, carried out in the 1930s. The investigators claimed to demonstrate the importance of taking into account the psychological needs of the individual worker in designing production tasks. These claims led directly to the general 'human relations' approach widely adopted since then in industry. It has been demonstrated, however, that the individuals on which the studies were carried out displayed particular characteristics that invalidated much of the supposed scientific nature of the whole investigation. In particular, Alex Carey has shown that in the first of the set of five studies, two of the original five girls selected for observation were dismissed for 'gross subordination'. They were replaced by two other girls, one of whom immediately took charge of the discipline of the whole group and led a sustained acceleration of output. In the light of this and similar evidence, Carey suggests that 'the limitations of the Hawthorne studies clearly renders them incapable of yielding serious support for any sort of generalization whatever'. He suggests that the results, 'far from supporting the various components of the "human relations approach", are surprisingly consistent with a rather old-world view about the nature of monetary incentives, driving leadership and discipline.'[44] So much for the neutrality of social science.

An important management technique reflected in production technology is the use of the hierarchical tree as a means of organizing production. This is a form of organization by which groups at every level are made subordinate to a single individual at the level above. The inverted tree provides the ideal organizational pattern by which information can flow up to management, and instructions down to the work-force. As in the army, political power and the possibility of control is fragmented at the bottom and consolidated at the top. The desire to maintain hierarchical relationships has led to the adoption of production techniques that appear to require them. The design of assembly-line production techniques, for example, coincides with a particular mode of the division of labour. As Gorz has written, 'hierarchical regimentation . . . *appears* to be a necessity that flows from production technology; but in truth it is built into production technology insofar as the latter is itself a reflection of the social division of labour.'[45] Again we see technology reflecting and reinforcing social relationships.

A further way in which technological innovation is used to manipulate the relationship between capital and labour lies in the activity of productivity bargaining. This is often associated with the introduction of innovations, being offered under the sugar-coating of higher wages to make up for possible unemployment created by the innovation. Productivity deals, according to two industrial sociologists, Bob Sutcliffe and Andrew Glyn, 'involve an effort by management to restore some *lost control* over work or pay, and therefore assist profit margins'[46] (my italics). The way these are linked with innovation, and the position of the workers concerned, are illustrated by a major strike at Rolls Royce in 1969. The strike cost the AEUW, one of the major unions involved, almost £¼m. At issue was the negotiation of a productivity agreement which demanded 'the acceptance of shift-work in order to exploit high capital equipment, the acceptance of work management techniques, the division of work into basic elements and the setting of time limits for these elements, such time to be compared with actual performance'.[47] The spread of capital equipment had led to the imposition on technical workers of the same methods of subordination and discipline that the machine has already brought to those on the shop floor.

When we talk of modern technology, we must of course take into account more than the technology of industrial production. Equally important is what might be called domestic or consumer technology. This includes all the various machines and gadgets, from cars to electric tooth-brushes, that form part of our techno-logical environment. In the way that industrial technology reflects the mode of social production of a society, consumer technology reflects the instrumental actions of its individual members. Yet to the extent that consumer technology is a product of industrial technology, and furthermore that it is based on iden-tical concepts of technological rationality – at least in its purely *functional* aspects – the one can be considered as being an extension of the other. Marx realized early on that patterns of consumption in general are no more than extensions and reflec-tions of patterns of production. He wrote in the *Grundrisse*, 'Production produces consumption; first by furnishing the latter with material; second by determining the manner of consumption; third by creating in consumers a want for its products as objects of consumption. It thus produces the object, the manner and the desire for consumption.'[48]

These processes have become increasingly apparent since the time of Marx. Industry, having provided for the basic needs of society, has moved on to creating new needs in order to maintain its expansion. Higher wages have ensured that the workers would be able to buy these new products, achieving satisfaction by ap-pearing to reach their 'social expectations'.

We are frequently reminded that we live in a consumer society, in which the tastes and apparent needs of the community are manipulated by advertising techniques so as to ensure a secure market for industrial products. According to Nathan Rosenberg, even in nineteenth-century America, consumer innovations were already greatly aided by 'the malleability of consumers' tastes . . . The public's willingness to accept a homogeneous final product was indispensable in the transition from a labour-intensive handi-craft technology to highly specialized machines'.[49] It is important to realize the extent to which this manipulation was built into the products themselves. This is neatly summed up by the remark attributed to Henry Ford: 'You can choose any colour car you want, so long as it's black.' There may be more colours now, but the principle is still there: by creating *apparent* choice in relatively unimportant items such as the texture of car seats or optional cigarette lighters, the *real* choices, such as price, mechanical reliability or ecological impact, are obscured and made to appear relatively insignificant.

Another example is provided by the development of television. While appearing to democratize society by disseminating information to all, and 'opening up the insides of politics', what we actually see on television is subject to strict centralized control experienced through imposed norms of acceptable viewing. The important choices – such as *how* news is presented, or *which* experts are invited to discuss important topics – lie outside the control of the viewer. Choices are limited to the colour and size of the set, and which of a severely restricted range of programmes to watch. The important question to ask about television, as of any major technological innovation, is not how it could be ideally used, but why it is used in the way that it is, and how does this usage become reflected in the design of the technology of communication as a social institution?

A further example of manipulation of consumer choice is the concept of 'built-in obsolescence'. Many products are designed to fail after a certain length of time purely in order to stimulate the flow of market commodities. This ensures a permanent market for the products of industrial research itself determined by the economic cycle of innovation and obsolescence. Other examples could be used to show how the nature of the product itself assists the manipulation of consumer demand in both the economic and political interests of a dominant social class.

The political nature of technology is not confined to man's social environment. The natural environment, too, has suffered under the same process. We have already seen how nature plays an integral part in the lives of most tribal societies, a presence to be treated with respect and, often, fear. The technologies developed within this framework reflected these general relationships. In the Middle Ages, the system began to break down on two fronts. On the practical side, the development of agricultural techniques enabled man to obtain greater productivity from the soil, inevitably turning him into the role of exploiter. This could only be justified, however, once the more mystical notions of nature had been swept away. It was here that the Church came to the rescue. The Judaeo-Christian tradition had always preached that man was the centre of God's universe – both literally as well as metaphorically – and that the forces of nature were there for his use. Nature remained personified in the form of the goddess Mother Nature. The general orthodoxy of the Western medieval world, however, was that God was the first absolute, and Nature his first minister and deputy.

The distinction between Nature and God coincided in many ways with the ideological or philosophical separation between man and nature. Raymond Williams has suggested it was this distinction that eventually made it possible to describe natural processes in objective terms. Man was now prepared, not only to observe nature, but to experiment with it in a scientific manner, and this included the conscious intervention by technological means for human purposes. 'Agricultural improvement, the industrial revolution,' he said in a lecture at the Institute of Contemporary Arts in 1970, 'follow clearly from this emphasis, and many of the practical effects depended on seeing nature quite clearly, and even coldly, as a set of objects, on which man could operate.'

Mumford has expressed it slightly differently: 'The machine came forward as the new demiurge that was to create a new heaven and a new earth.'[50] He goes on to quote Agricola's description of the environmental havoc created by mining techniques in Italy as early as the sixteenth century. Critics of these techniques, according to Agricola, argued that 'woods and groves are cut down, for there is need of endless amount of wood for timbers, machines, and the smelting of metals. And when the woods and groves are felled, there are exterminate the beasts and birds, very many of which furnish pleasant and agreeable food for man. Further, when the ores are washed, the water which has been used poisons the brooks and streams, and either destroys the fish or drives them away.'[51]

It seems important to point out, particularly today when we are told that we are in the middle of a world-wide ecological crisis, that environmental despoliation is a predictable consequence of our current patterns of economic and technological development. As we have already mentioned, environmental factors are listed under the external costs of production, namely costs to be shared by the community as a whole, rather than under the internal costs born by the productive enterprise which causes the pollution. The role of the efficient manager is to minimize the internal costs of the production process and hence by implication – in a case where the combined costs can be taken as constant – to maximize the external costs within this overall constraint.

Galbraith points out that 'beyond the area of goods and services, however supplied, and the demand for them, is the further world of aesthetic experience'.[52] Environmental considerations of a non-economic nature come into this category, which Galbraith uses in a far broader sense than merely the fine arts. He adds that 'aesthetic achievement is beyond the reach of the industrial system and, in substantial measure, in conflict with it. There would be little need to stress the conflict were it not part of the litany of the industrial system that none exists'.[53]

Of course, technological innovation, either by increasing the productive efficiency of the extraction process, or by creating substitutes that will remove the pressure from scarce natural resources, *can* result in a net decrease in environmental exploitation in particular instances. Similarly legislative constraints can act to reduce critical levels of environmental damage. The main point, however, is that a basic law of the capitalist market economy still applies – the maximum creation of surplus value. The price of raw materials needed for industrial production depends on the cost and efficiency of extraction and transport processes (including labour costs), the amount paid to central licensing authorities, and the *economic* consequences of possible scarcity. It does not depend on the amount of damage inflicted on the environment *per se*, which can thus be maximized within existing technical, economic and legislative constraints. Even the declared need to preserve scarce resources for future generations is the expression of an economic, and not an environmental requirement.

In a situation, admittedly hypothetical, in which all other costs are kept constant, modern capitalist technology demands the maximum exploitation of available environmental resources, as these start out as almost pure surplus value. Herbert Marcuse goes so far as to lay the blame for this whole process on the framework of 'scientific rationality' that lies at the heart of modern technology. 'Science, by virtue of its own methods and concepts,' he writes, 'has projected and promoted a universe in which the domination of nature has remained linked to the domination of man – a link which tends to be fatal to this universe as a whole.'[54]

In capitalist society, direct social control was removed from the alleged representatives of the community by the early capitalists during the first stages of the Industrial Revolution. It is now apparently being transferred back in some countries to the social body. In the UK, for example, this has been done through the nationalization of key industries, the intervention of Government in all sectors of the economy, and, more directly, by measures such as the Industrial Relations Act and the prices and incomes control. Such moves, however, only extend the above analysis of the process and incentives of technological innovation from the private into the public domain. Government is no less willing than industry to use technological innovation as a powerful political tool. In a recent lecture sponsored by the *Times Literary Supplement*, Lord Zuckerman, for many years chief scientific adviser to the British Government, put the position clearly. 'In the national interest', he said, 'the application of the results of scientific enquiry must necessarily be biased by political and administrative considerations.'[55] One of these considerations must be the apparent need of the main decision-making bodies to control the activities of all members of the community, justified in terms of the need for 'law and order'.

Perhaps the most direct manifestation of the political nature of technological innovation is military technology. This is specifically designed to provide the maximum control over opposing political forces. Recent examples of this have been highlighted by the development of weapons of chemical and electronic warfare, and the rapid rate of technological innovation in other more conventional types of weapons by the US for use in Indochina. Steven Rose, professor of biology at the Open University, has written that 'these applications of science are not neutral, neither are they inevitable, nor are they accidental; they are the result of a set of deliberate and specific choices about the types of weaponry and the cost-effectiveness of their use'.[56] If we realize that the term 'application of science' embraces, not only military weapons, but also the set of decisions leading to the development and use of these weapons, we again see the way in which dominant social values become embedded in the technology that they produce. The arguments relating to military technology can be extended to the whole technological base of modern society. In the previous chapter, it was shown how the conventional history of technology is written in terms that imply it to be determined by the functional interdependence of technological and economic development. This appears to be based on the desire to increase the technical efficiency of the productive process, and thus to maximize the creation of surplus value. We now see that the history of technology can also be interpreted in social and political terms. From this perspective, it can be seen as the manifestation of the needs of a dominant social class – reflected in apparent economic need – to maximize the control over the activities of the work-force, and the exploitation of natural resources.

Mumford concludes at the end of his *Technics and Civilisation* that 'the Western European conceived the machines because he wanted regularity, order, certainty, because he wished to reduce the movements of his fellows as well as the behaviour of the environment to a more definite, calculable basis'.[57] The objective necessity for doing this was contained in the direct expression of economic and class interests. Developed as a coherent system to meet both the economic and the political needs of these interests, technology as a social institution necessarily came to incorporate these needs. To use a possible metaphor, technology was the key that unlocked the door to 'progress'. A key reflects the design of the lock it has been made to open; to open a different door requires a different-shaped key. A technology primarily developed under authoritarian capitalism both reflects and becomes part of an ideology that permits the exploitation of man by man, and the destruction of the environment in the 'interests' of society. An alternative pattern of social development seems inevitably to imply an alternative technology.

We now see why it is necessary to go beyond merely technical reasons to understand the social role played by technology. Although maximum productivity requires strong centralized control, capitalism prefers to make this control normative and ideological rather than overtly authoritarian and coercive. In his *Eros and Civilisation*, Marcuse develops Freud's ideas on the psychological problems of the individual in a technological society. Marcuse distinguishes a factor which he labels 'surplus repression' which he claims to be the characteristic of all such societies. He writes 'Domination differs from rational exercise of authority. The latter, which is inherent in any social division of labour, is derived from knowledge and confined to the administration of functions and arrangements necessary for the advancement of the whole. In contrast, domination is exercised by a particular group or individual in order to sustain and enhance itself in a privileged position.'[58]

Marcuse suggests that the pleasure principle defined by Freud was 'dethroned' not only because it militated against progress in civilization, but more importantly because it opposed a civilization whose progress perpetuated domination and toil, the progress of capitalist society. The ideology of industrialization, which implies the neutrality of technological development, masks the nature of this domination and cements the mechanisms of political control.

Contemporary technology has been developed in industrialized societies in a way that seeks to secure maximum control over labour, as much as the maximum production of goods. Indeed, when control and productivity conflict, it is often productivity that is sacrificed. The modes of production have therefore been developed in a way that both supports and legitimates the existing relations of production, and the social attitudes by which these are expressed. When Harold Wilson offers sympathy to the individual dominated by what appear to be the oppressive forces of technology, he is in fact doing no more than describing the *political* oppression experienced by the individual. He is using an ideological disguise to pretend that the problem is a technical one that can be solved by increasing the effectiveness of our existing social institutions – in this case, by putting a Labour Government back into power.

The political perspective indicated above helps us to make sense of the social problems associated with contemporary technology described in the first chapter. Every so often the ideological disguise slips and the contradictions it covers stand nakedly revealed. Society may overstep the mark in its attempts to stretch the conditions of ecological equilibrium to its limits, committing irreversible damage to the environment; as it has often done in the underdeveloped countries, industrialization may fail to deliver the goods that it has promised, and by doing so, invite criticism and analysis of its 'inappropriateness'; the efforts of management to alleviate the alienation of its work-force may be inadequate to mask the social problems created by a particular work-situation, resulting in psychological disturbance that can be objectively measured. These and other such problems, like the cracks made in a prison wall, reveal – if we are prepared to look through them – the nature of the activities that lie behind them. They also give an indication of the extent of their ideological disguise, which represents the social problems associated with advanced technology as isolated problems accessible to technical solutions.

What are the alternatives? Given that technological development is a political process, and that the technology we possess at present coincides with the authoritarian and hierarchical relationships of capitalist society, is it possible to conceive of a technology that is based on non-authoritarian, non-hierarchical relationships? Such a technology should, ideally, provide the framework for the expression of collective initiative and community control, as well as the development of the full capabilities and creativity of the individual. Is it possible to create a society in which the distinction between social and technological values, the first reflecting the values of man, the second those of the machine, no longer exists? This chapter has indicated how technological change must be viewed as a political process, reinforcing the interests of a dominant social class. It also implies that development of a non-alienating, non-exploitative technology requires more than just a nominal change in ownership of the machines we have now. They include a complete reshaping of our attitudes towards the function of technology in society – a simultaneous change, in other words, of both political and technological consciousness.

4 Utopian Technology:

some basic principles

We are now in a position to discuss in greater detail the concept of alternative technology. In the previous two chapters, we have seen how industrialization is portrayed as a technical and objective necessity for social development. The birth of industrialization during the Industrial Revolution, however, can alternatively be seen as a political event: the technically articulated response to the economic and political determinants of industrial capitalism. The same is true of the accompanying development of the factory system. Given a different set of determinants – in other words a different set of cultural values, of ownership relations, and of social priorities in the broadest sense – is it possible to contemplate an alternative form of social development to that which has resulted in large-scale, centralized patterns of industrialization, and the alienated life-style that has come with it? The tools and machines required to maintain this alternative would necessarily embody a very different set of social and cultural values from those we possess at present. These tools and machines, together with the techniques by which they are used, form what is generally meant by the term alternative technology.[1]

The situation can be stated more explicitly. The Industrial Revolution resulted largely from the coming-together in eighteenth-century England of three sets of skills and techniques: the entrepreneurial skills required to create and exploit market opportunities for the products of the early factories, the management skills needed to handle the organizational aspects of the factories, and the technical skills that provided the means of increasing output and maintaining the expansion of the whole economic machine. These skills shared a common ideological base, one that gave social expression to a concept of rationality that placed emphasis on individualism on the one hand, and the maintenance of hierarchical and authoritarian forms of social control, as well as a dominating and exploitative attitude towards the natural environment, on the other.

The problem, of course, faced by those involved with industrial development at this time was that each set of skills or techniques had to be developed on a largely empirical basis. Invention was necessary in all fields and became encouraged by the spirit of 'rational' inquiry inherited from scientists and natural philosophers such as Galileo, Descartes and Francis Bacon. This process, as we have previously indicated, played a crucial part in the transformation from a 'closed' society emphasizing the cohesiveness of all aspects of social activity, to an apparently 'open' society based on ideas of expansion, and continually seeking innovation and novelty.[2]

Today we are in a very different position from that faced by the early industrialists. The knowledge of the natural world derived by science, and our historical experiences of technology, have given us a view of the potentialities for social development that was inconceivable within the limited perspective of eighteenth-century England. In particular, we have seen the problems resulting from an irrational belief in the innate powers of industrialization – whether applied by a political system that is nominally socialist or capitalist. We should now be in a position to exercise a sensible choice over the managerial, entrepreneurial and productive techniques necessary for social development compatible with our ideas of the nature of man and his role in society. Our knowledge and experience should enable us to displace the dominant ideology and the social practices on which it is based by one that ensures technological innovation remains responsive to direct social needs, and maintains a non-alienating and non-exploitative relationship between man and man.

The purpose of this and the following chapter is to describe some of the tools, machines and techniques that we might choose if faced with the possibilities offered by this situation. Much discussion of alternative technology tends to concentrate on eco-logical aspects. It stresses, for example, the need to develop technologies that make minimum use of non-renewable resources, that do not pollute the environment or present a hazard to the health of the community, and in general that cause as little interference with natural ecological cycles as possible. Such a technology would be based on a sense of co-operation, rather than competition and domination, between man and nature. Thus sections of these two chapters deal with alternative sources of energy – such as wind-power, solar energy and water-power – alternative techniques of building, of growing food, and of providing for general community needs in a way compatible with this attitude towards the natural environment.

Equally important, however, are the ways that an alternative technology might relate to the non-material needs of the individual, and of the society or community of which he is a member. To avoid the problems of alienation, for example, inherent in social production as it is carried out in capitalist society, we have seen that it is important to change not only the ownership of the machines, but also those institutionalized aspects that embody the authoritarian and hierarchical relations of production of industrial capitalism. To put it more simply, we must do away with machines that appear to treat men as mere appendages, and replace them with machines that the individual can operate in a way that is both socially productive and personally fulfilling.

Of course, much of this is a personal matter. Different people will obtain satisfaction from working with different types of tools or machines. It is therefore difficult to write down a list of machines which can be labelled as 'non-alienating', in particular as this would imply the separation of a machine from its social environment. All one can do at this stage is to suggest some principles by which these technologies might be selected. For example, since many of the problems associated with industrialization result directly from its large-scale, centralized nature, it might be suggested that an alternative form of social production would be one based on small-scale decentralized units, each under the control of both those concerned with the production process, and those who will use the services or commodities that are produced.

This immediately brings us back to the political dimension. It is not enough to say that industrialization is an ideology without taking into account the political structure that the ideology hides. We have emphasized that a society's technology embodies the structure of its productive activity, and of its social relations of production. This underlines the fact that the root causes of the problems associated with industrialization are essentially political as much as technological. An alternative technology can therefore only be successfully applied on a large scale once an alternative form of society has been created. The task of doing this is a political, rather than a technological one. The consequences of confusing these two aspects will be highlighted in a later chapter when the concept of intermediate technology as applied to the problems of the underdeveloped countries is discussed.

From the political perspective, any discussion of alternative technology must be necessarily utopian, being in Karl Mannheim's phrase, 'incongruous with the state of reality within which it occurs.'[3] In this chapter, therefore, the various tools, machines and techniques described will be given the label 'utopian technology', and those at present concerned with their development referred to as 'utopian technologists'. This is not done to imply that the tools and machines described are impracticable, but to indicate that their introduction on a significant scale would be virtually impossible within the existing structure of society. However it does make it necessary to refer briefly to the historical development of utopian thinking, and its relationship to ideas about science and technology.

The term 'utopia' was first used by Thomas More. He attempted to find a way of escaping the corruption of wealth and power that accompanied the rise of the new bourgeoisie in the early sixteenth century. He suggested this might be done through the creation of a form of communist society based on the equality of all individuals, derived partly from Plato and partly from early Christianity. Both More and, subsequently, Francis Bacon emphasized the crucial role to be played by science in any ideal society. Bacon's imaginary House of Solomon in his own utopia, New Atlantis, was a kind of universal laboratory. Elsewhere he argues, referring to the dominance of the Faustian tradition, that whatever knowledge might be forbidden to man, it is not natural science.

Of the later socialist utopians, the three most important were Saint Simon, who envisioned a society in which scientists, artists and artisans would 'direct the work of the nation'; Charles Fourier, whose curious mixture of radical and reactionary ideas provided a remorseless critique of the 'material and moral misery' of early industrialization, demanding the abolition of factories and a return to a semi-rural society; and Robert Owen, who tried to operate his New Lanark factory along socialist lines, taking into his scheme all aspects of social life such as housing and education, yet found himself ostracized from Victorian society as soon as the full political implications of his ideas were realised.

In virtually all early utopias – except possibly those of Fourier and later William Morris – technology played a relatively neutral role. The main emphasis was placed on a change in life-style and social organization; the technological problems involved were left largely unresolved, and the choice of techniques to reach particular objectives left as a technical matter. What distinguishes many utopian technologists from earlier utopians is that, having realized the non-neutrality of technology, they consciously seek to design the life-style, and the technology that would go with it, as an integrated whole.

Utopian technology remains based on a set of ideals. The implications of its introduction present a serious challenge to the present organization of society and its existing technologies. At the same time, however, these ideals are firmly placed in technological practices that, in many cases, have already been successfully developed, albeit in a piecemeal and fragmented fashion.

The strong emphasis on practicability and the material aspects of utopianism again distinguishes utopian technologists from many of their earlier antecedents whose ideas have usually been on a more idealistic and almost metaphysical level. By combating idealism in this way, the proposals of utopian technologists are closely allied to the idea of a classless society put forward by Marx and Engels in the *Communist Manifesto*, although without the emphasis that the latter place on centralization of political control and the extension of the factory system of production. Utopian technologists imply the need for a re-interpretation of historical materialism that challenges the objective character of the contradiction between the modes and the relations of production, claiming the two form a political unity, and must be combated together.

The general criteria according to which utopian technology is designed are those intended to avoid the type of problems discussed in previous chapters. It forms a framework that is designed to eliminate the alienation and exploitation of the individual, and the domination of the environment by the activities of man. Utopian technology might be characterized as a technology which is satisfying to work with, can be controlled by both the producers and the community by whom the products are used, conserves natural resources, and does negligible damage to the environment. Another characterization might be as 'a low benefit, low-risk technology': although the general standard of living – as measured in terms of quantities of material goods, etc. – is likely to be considerably lower than that provided by contemporary industrialized societies, so too are the risks that utopian technology presents to the social and environmental fabric.

In more precise terms, John Todd of the New Alchemy Institute in the US has selected five characteristics of what he refers to as 'biotechnology'. His suggestions are: that it should function most effectively at the lowest levels of society; that the 'poorest people' should be able to use it; that it should be based primarily on ecological and social considerations, rather than those of economic efficiency; that it should allow the possible evolution of small, decentralized communities; and that it should require relatively small amounts of resources.[4] Similarly Peter van Dresser, who has been developing similar ideas in New Mexico for twenty years or so, suggests that the renewal of smaller rural and provincial communities should be based on 'skilled, scientific and conservative use and management of local biotic and other flow resources, rather than on large-scale machine and energy-intensive industries and heavy transport.'[5] Colin Moorcraft, in an issue of *Architectural Design* devoted to alternative technologies, describes the same general ideas slightly differently. 'The basic aims of any new systems of man-environment interactions,' he writes, 'must be the restoration and maintenance of the overall entropy-minimizing tendency of the planet – i.e. earth must be brought back to life and stay healthy.'[6]

Three points emerge from attempts to lay down criteria for the design of utopian technology. The first is a certain lack of precision. There are no ready-made formulae available for the purely mechanical design of utopian technology. The second point is the extent to which technical, social, environmental and political criteria are integrated into one another. Rather than emerging with a neat list of independent criteria, we find one criterion depends on the others to an extent that makes fragmentation of function meaningless. Finally, a common theme is the need to get away from a primarily economic approach to the evaluation of human activity and the use of environmental resources. It is possibly here that the most serious challenge to the existing economic system – and hence the most radical political implications – are to be found. Capitalism, as Marx demonstrated, is founded on the concept of surplus value, which expresses the value of both individuals and material goods in economic terms; it is this very fact that distinguishes capitalism, according to Marx, from all earlier forms of social organization (and will indeed lead to its eventual destruction). When *A Blueprint for Survival* points out in a slightly surprised way that ' "economic value" as at present calculated does not correspond to real value any more than "economic cost" corresponds to real cost', it is saying no more than Marx did over one hundred years ago when he distinguished between the use-value of a commodity – which is 'only realized in use or consumption' – and its exchange value in a market economy. To challenge the validity of this distinction is thus implicitly to challenge both the economic and the political basis of the capitalist system.

A detailed list of criteria for what he calls 'soft technology' has been prepared by Robin Clarke of Biotechnic Research and Development (BRAD). This gives a clear indication, admittedly in a somewhat idealistic form, of the type of characteristics that might distinguish utopian technology from the advanced technology that we have at present (see Table 1). Again the interdependence of many of these characteristics is immediately apparent, for example the concepts of being 'ecologically sound', 'integrated with nature' and having 'technical boundaries set by nature'. Others, although desirable, might in practice turn out to be contradictory, such as demands for cheapness and low specialization, and require some form of compromise. However, what is important for our present purposes is the general approach indicated by these characteristics. They form a coherent system that can only be interpreted as a whole, and loses much of its sense when reduced to fragmented components.

SOME UTOPIAN CHARACTERISTICS OF SOFT TECHNOLOGY
(Robin Clarke)

'Hard' technology society	*'Soft' technology society*
1 ecologically unsound	ecologically sound
2 large energy input	small energy input
3 high pollution rate	low or no pollution rate
4 non-reversible use of materials and energy sources	reversible materials and energy sources only
5 functional for limited time only	functional for all time
6 mass production	craft industry
7 high specialization	low specialization
8 nuclear family	communal units
9 city emphasis	village emphasis
10 alienation from nature	integration with nature
11 consensus politics	democratic politics
12 technical boundaries set by wealth	technical boundaries set by nature
13 world-wide trade	local bartering
14 destructive of local culture	compatible with local culture
15 technology liable to misuse	safeguards against misuse
16 highly destructive to other species	dependent on well-being of other species
17 innovation regulated by profit and war	innovation regulated by need
18 growth-oriented economy	steady-state economy
19 capital intensive	labour intensive
20 alienates young and old	integrates young and old
21 centralist	decentralist
22 general efficiency increases with size	general efficiency increases with smallness
23 operating modes too complicated for general comprehension	operating modes understandable by all
24 technological accidents frequent and serious	technological accidents few and unimportant
25 singular solutions to technical and social problems	diverse solutions to technical and social problems

26 agricultural emphasis on mono-culture	agricultural emphasis on diversity
27 quantity criteria highly valued	quality criteria highly valued
28 food production specialized industry	food production shared by all
29 work undertaken primarily for income	work undertaken primarily for satisfaction
30 small units totally dependent on others	small units self-sufficient
31 science and technology alienated from culture	science and technology integrated with culture
32 science and technology performed by specialist élites	science and technology performed by all
33 strong work/leisure distinction	weak or non-existent work/leisure distinction
34 high unemployment	(concept not valid)
35 technical goals valid for only a small proportion of the globe for a finite time	technical goals valid 'for all men for all time'

Certain technologies, such as simple agricultural or gardening tools, can immediately be included as 'desirable'. At the other extreme, products of advanced technology such as colour-television sets and supersonic aircraft would almost inevitably be excluded. So, too, would be a number of technical developments that, while providing an adequate solution to a *particular* problem, nevertheless offend a sufficient number of criteria to be judged undesirable. Examples of this might be the gigantic satellites, up to five miles square, which some suggest could be put into orbit around the earth to capture large amounts of solar energy for use on earth.[7] Less easy to classify are technologies that, although requiring advanced techniques in their production, still provide a number of useful functions. The general use of plastics – light, cheap and durable, but resource-consuming and relatively difficult to manufacture – is one example of the dilemma faced by many utopian technologists. Others are split over the use of advanced electronics, for example video-recorders and computers. Again, both of these greatly increase the potential and, some feel, the whole viability of commune-living. Yet they offend a number of characteristics such as ease of maintenance generally associated with utopian technology. These dilemmas indicate a central difficulty running through the discussion of utopian technology that follows. It would be wrong to imply that utopian technology is a well-defined category of machines and tools. There are many different viewpoints on what should, or should not, be included in this category. What I hope to do is provide some examples of specific tools, machines and techniques that have been developed by particular individuals and groups, hoping to indicate the *types* of alternative approaches to technological development implied by the 'utopian technology' label.

There are three sets of relationships involved when we discuss the principles of utopian technology. These are the relationship of technology to the individual, its relationship to the community, and its relationship to the environment. From the point of view of the individual, utopian technology would ideally provide the means by which he could fufil and experience his full human potential. It would be non-alienating, in the sense of remaining directly under the individual's control, and linking him with, rather than separating him from, fellow members of the community, and the natural environment. Murray Bookchin writes in his essay *Towards a Liberatory Technology* that 'labour is not merely expenditure of energy; it is also the personalized work of a man whose activities are sensuously directed towards preparing his product, fashioning it, and finally decorating it for human use . . . the tool amplifies the powers of the craftsman as a *human*'.[8] Utopian technologies would seek to reflect this, stressing the close relationship between the individual, the tools with which he works, and the object that he produces. Even if productive activity was carried out in groups, the need would be emphasized to design tools and machines that reflect the needs and potentialities of the individual, an approach closely resembling that of traditional craft techniques. It would attempt to ensure that the machine remained an appendage to man, rather than the other way round.

The system of community values embedded in utopian technology usually includes some expression of the need to maintain the freedom of the individual from exploitation and control by others. At the same time it underlines the importance of communities being bound together by collective experience and ideals that enable the desires and needs of all its members to be met. The implications for technology is that it should be accessible to control by the community, and should not provide a means of exploitation or domination of its members. This could, for example, mean an emphasis on small-scale, workshop-size production units of no more than a dozen or so individuals, rather than the massive factories employing many thousands which now make up our industrial landscape. Industrial sociologists have discovered that many of the problems of alienation associated with production-line work can be alleviated by dividing the work-force into such small groups, giving each group responsibility for the construction of all aspects of a particular commodity, such as a car or a television set, the idea is often referred to as 'job enrichment'.[9] Utopian technology would also emphasize the interrelatedness of all aspects of social life. According to one group of British architects who have drawn up plans for a community that integrates all the different aspects of social activity, 'the most effective and least alienating means of producing what the community needs are within the community itself'. They add that 'the work situation should become a part of a living culture through its own organization, and emphasize that work cannot be divorced from other individual and community activities, but must be integrated into them'.

Of course, it may be neither possible nor desirable to cater for all the needs of the community on a local, workshop basis. There might still be a case for some centrally-produced products. It has been suggested, for example, that social production might be organized at three levels: neighbourhood workshops, in which craft-work could be redeveloped at various levels of technology; small multi-purpose factories able to create a variety of products by flexible tooling; and medium-size largely automatic factories at a regional level, producing both finished goods and standardized materials and parts for community factories and workshops.[10] Other supporters of utopian technology might disagree on whether automation was a desirable feature. All seem to agree, however, on the need for a high degree of control over all aspects of production exercised by both the community and the workers involved.

What is needed,' writes Paul Goodman in his book *Communitas*, 'is the organization of economic democracy on the basis of the productive units, where each unit, relying on its own expertness and bargaining power of what it has to offer, co-operated with the whole of society.'[11] He suggests, as the group of architects quoted above do, that there should in general be a closer relation between 'personal' and 'production' environments.

It is difficult to be more specific about the exact nature of the type of tools and machines that might be developed along the above lines. This is largely due to the relevance of such issues to the *relations* between these machines implied by the way they are used, rather than the nature of the machines themselves. In broad terms, the machines would seek to 're-humanize' the content of work. This might involve the revival of a craft approach by which work becomes once again task-oriented rather than wage-oriented. It would also imply the reintegration of production tasks that would avoid the alienation accompanying some of the more extreme examples of the division and fragmentation of labour. Reducing the scale and complexity of industrial plant would furthermore help alleviate the alienation of those who now feel removed from control over large-scale processes in which they now appear to perform a relatively insignificant role. Similarly, control of design and innovation procedures by the workers involved might help towards the development of non-alienating processes. The whole concept of self-management, however, implies that beyond indicating general principles, one should not try to lay down in greater detail the type of production technology that might be adopted in a particular instance. Such decisions would be left essentially to the workers and the community involved.[12]

It is possible, however, to be more definite about the third relationship, that between technology and the environment. Here we have a number of facts – such as potential levels of resource-usage and environmental interference – to work on. We also have a certain amount of knowledge about the ways in which the environment works, represented in the discipline of ecology. Starting from this situation, it is not too difficult to build up a framework of technologies that should, theoretically, minimize the

damage to the environment and help us to make use of availab
resources in the most socially and ecologically-desirable wa

Ecologists separate ecosystems into two functional componen
The autotrophic part captures the sun's energy and produces foo
from simple organic substances, while the heterotrophic pa
obtains its energy from the food that is produced, decomposing th
complex materials back into organic compounds. The who
biological chain of life of which we form a part depends on th
fixing and transfer of energy received from the sun, and the con
tinuous recycling and re-utilization of energy and materia
through the ecosystem. The biosphere, as it has come to be called
is a complex network of processes, each dependent on the othe
to an extent that forms it into an organic whole. In less technic
terms Professor Barry Commoner has formulated what he con
siders to be the four basic laws of ecology: everything is con
nected to everything else; everything must go somewhere; natu
knows best (i.e. interference with ecological systems can be equiv
alent to poking a pencil at random into the back of a watch); an
there's no such thing as a free lunch.[13]

Where do the activities of man stand in relation to natur
ecological processes? It is claimed to be possible to build comple
'thermodynamic' models by which the natural entropy-increasin
tendencies of the ecosystem (i.e. those which push it towar
states of ever-greater disorder) are offset by processes of evolutio
and natural selection. These are assisted by the activities of in
dustrial man who imposes a form of technological order on en
vironmental events, leading hopefully to a state of equilibrium
Such models, however, tend to be based as much on faith an
ideological prejudice as they are on sound scientific argumen
All we need say at present is that Commoner's laws of ecology a
openly ignored, if not actively broken, by advanced industrialize
societies, often with dramatic consequences. It is thus reasonab
to suggest that if man's activities were designed in a way tha
acknowledged the validity and importance of these laws, th
chances of creating widescale and irreversible environment
damage would be greatly reduced, and perhaps even eliminate

ENERGY

Let's start with energy. This lies at the heart of all ecological and technological systems. Nearly all the energy used on the earth comes from the sun. Since the temperature of the earth remains relatively stable (neglecting the variations that occur over long periods of time), it is possible to say that the amount of heat received from the sun in the form of radiation at any one time – about 1.5×10^{15} kilowatts an hour, or 35,000 times the total energy consumption of all human activity – must be roughly the same as that lost through radiation back into space.

There are three ways in which the sun's energy is captured by the earth. The first is the energy absorbed by the atmosphere and those parts of the earth covered by water. This is the energy behind the movements of wind and water. It causes water, for example, to evaporate into the atmosphere and fall back to earth again as rain, subsequently running off the land into the sea. It also provides the energy of the waves, and ocean currents (although not the tides which obtain most of their energy from the movements of the moon).

A large amount of the sun's energy is absorbed directly by the plants and animals which form the biosphere. Plants absorb this energy through the process of photosynthesis, using it to turn inorganic chemicals from the soil and the air into vegetable matter. This provides the food for both marine and terrestrial animals, part of the general life-cycle that we have described briefly above. Finally, vast quantities of energy are contained in the earth itself and its crust, known as the lithosphere. A large proportion of this originated when the earth broke away from the surface of the sun many millions of years ago, and is now present as thermal, chemical and nuclear energy. But energy is also stored in the form of fossil fuels such as coal and oil, which have resulted from the decay of organic materials. The energy in the lithosphere is non-renewable, as it represents energy which was either part of the earth at its origin, or – like the fossil fuels – has been built up over a long period of time. In neither case is there any way it can be replaced on a short timescale by the intervention of man.

It is from this third store of energy, and in particular from the fossil fuels, that most of the power consumed by industrial society is derived. The main forms are coal, gas and oil. Between them, these sources accounted for over ninety-five per cent of the energy consumed by the US economy in 1971, with roughly two per cent contributed by nuclear power and three per cent by hydropower. Overall, it has been estimated that thirty-seven per cent of the world's use of energy is derived from coal, thirty per cent from oil and fourteen per cent from natural gas. As we are becoming increasingly aware, however, there are finite limits to the amount of each of these resources, and the extent to which we can rely on them in the future, since the world's consumption of energy is now doubling approximately once every ten years.

The production of heat from fossil fuels also poses major problems of the disposal of waste heat. According to one power engineer, 'a modern industrial society can be viewed as a complex machine for degrading high-quality energy into waste heat while extracting the energy needed for creating an enormous catalogue of goods and services.'[14] For thermodynamic reasons, waste heat is an unavoidable result of obtaining energy from fossil fuels. It can present major environmental problems. Some of these, such as the disposal of heated water from power stations, may be relatively localized. However the waste heat from all man's energy-consuming activities is now also claimed to be having considerable effects on the weather and the general global climate.[15]

The utopian technology approach to energy production is based on the premise that we should be using non-renewable fuels as little as possible. Alternative sources are therefore required. These usually involve the direct use of solar energy, its indirect use through wind or water-power, or the use of renewable resources such as wood and methane gas. (Another alternative is the use of 'geothermal' energy – i.e. the heat of the earth's interior – available in the vicinity of hot springs and geysers, but this tends to be either of relatively localized importance, or to require high-power transmission over long distances.) The efficiency of energy conversion from such sources is relatively lower than from the standard fossil fuels, and particularly from nuclear energy. This is felt to be the price that must be paid, however, for developing a technology and a life-style that is designed to fit in with environmental considerations. A further advantage of power schemes which use the wind or solar power collected at the earth's surface is that, unlike the use of fossil fuels, they do not add any heat load to the earth's biosphere, since there is no waste heat involved in the primary energy conversion process.

SOLAR ENERGY

For countries with a high level of sunshine, the direct conversion of solar energy appears the most significant long-term alternative to the use of either fossil or nuclear fuels. Interest in the use of solar energy, on both a large and a small scale, has been increasing rapidly during the past few years. Although primarily of interest to those living in tropical climates, there are nevertheless a number of ways in which the sun's energy can be put to productive use in some countries as far north as sixty degrees latitude, roughly the level of the Shetland Islands. It has been estimated, for example, that large areas of the United States could obtain much of their energy requirements from the sun. 'If only a few per cent of the land area of the US could be used to absorb solar radiation effectively (at, say, a little better than ten per cent efficiency), we would meet most of our energy needs in the year 2000,' the US engineer, Chauncey Starr has claimed.[16]

A study carried out by the Department of Mechanical Engineering at the University of Maryland suggested that for most of the US solar energy is already competitive with other forms of energy for space and water heating over much of the year. A report by an expert panel of the US National Science Foundation and the National Aeronautics and Space Administration recommended a three and a half billion dollar research programme over ten years. This would investigate the various possible uses of solar energy, and the National Science Foundation has already agreed to allocate five million dollars in 1973 to research in this area. In countries such as the British Isles with lower amounts of annual sunshine, the potential uses of solar energy are correspondingly less, but are still far from negligible.

Solar energy can be used in two ways. It can be converted directly into heat to provide heating for a house – an extension of the green-house principle. Alternatively the energy can be used indirectly to generate a source of electrical power. This can either be done using solar cells, similar to those widely used to power satellites, or through producing extremely high temperatures by focusing large amounts of sunlight on to a single spot, and then using this heat to power a generator.

The simplest way to capture solar energy is to let the radiation fall on a black surface. Depending on the nature of the surface, this will absorb almost all of the incident radiant energy and turn it into heat. As soon as the temperature of the black surface rises above that of the surrounding environment, however, the surface itself begins to lose heat through radiation. There are three main ways of preventing this from happening. The first is to remove the heat from the surface as quickly as possible. This is usually done by passing air or water across the surface to pick up the heat and transfer it to wherever it is being used or stored. The second is to insulate the surface in some way to prevent the heat from escaping. A typical way of doing this is to place a glass sheet slightly above the black surface, letting the sun's radiation pass through one way, but preventing the heat from escaping out again (the principle on which a green-house works). Finally, heat loss can be reduced by focusing the sun's rays on to a single small absorptive surface. Heat is again conducted away from the surface in some way, but greater efficiency can be achieved by this process since it works at higher temperatures than the other two methods.

The first two of these techniques have been widely used in designs for simple water-heaters. Perhaps the simplest example is the flat-plate heater. This is typically a flat glass plate with a black heat-absorptive surface a few inches behind it, tilted to face the sun. Maximum exposure is obtained from a tilt of ten to fifteen degrees plus the latitude at which the heater is being used. In a closed, siphon system, cold water enters the gap between the glass plate and the black surface at the bottom, rises as it is heated, and is finally run off from the top into a storage tank. Alternatively, it is allowed to fall over the plate under its own weight, and is taken off from the bottom. It is then either drawn off the tank as required, or returned to the surface of the plate again for further heating. A flat-plate collector built by two students at the Thames Polytechnic in London, J. C. McVeigh and Graham Caine, was found to raise the temperature of water in a tank on a sunny afternoon in January to almost twenty degrees Centigrade, while water in a neighbouring unheated tank remained at eight degrees Centigrade. Temperatures as high as sixty degrees were recorded in the tank in April. Such heaters, however, as with all forms of solar energy, depend greatly on prevailing weather conditions. The two designers have openly admitted that the high costs involved in producing substantial amounts of energy from flat-plate collectors in the British climate severely affects their viability.

A more promising area for making direct use of solar energy is in its potential for heating houses. Here there are three main problems to be solved. The first is the collection of the solar energy. This is usually done through the use of large absorptive surfaces similar in design to the flat-plate collectors described above. The second major problem is that of storage; unless some auxiliary form of heating is used, it is usually necessary to arrange for storage to make heat available, both for use during the night, and for those days on which the sun is not shining. Finally, there is the problem of distributing the heat through the house. Here the most commonly-used method is the circulation of hot-air through ducts in the floorboards.

One of the earliest attempts at solar heating was the Dover House, designed by Dr Maria Telkes and Eleanor Raymond and built in 1949. In this house, energy from the sun is absorbed by a large area of blackened metal sheets covered by double plates of glass. The heat is carried away by air circulating behind the metal sheets. It is stored chemically in large tanks containing Glauber's salt (sodium sulphate decahydrate), a given volume of which can hold eight and a half times more heat than water, if its temperature is kept between twenty-five and thirty-seven degrees centigrade. The house is heated by hot air conducted from the storage tanks by a fan (itself electrically driven). Dr Telkes' ideas are now at the commercial stage, and are claimed to be applicable to the houses of two-thirds of those living in the United States.

A variation on the same principle is the famous solar roof constructed by Harry E. Thomason on his house in Washington, DC., in 1959. He has covered an ordinary roof with a layer of insulation, then with blackened sheets of corrugated aluminium, and finally with 840 square feet of glass. Heat is conducted away from the roof by water which runs down the gulleys in the aluminium and is collected in the basement of the house. Here the heat is transferred to large bins containing fifty tons of small rocks. Again fans are used to circulate hot air around the house. Harry Thomason claims to get sixty to ninety per cent of house heating from solar energy in the winter. The sun also provides him with all his hot water, as well as cooling the house in the summer.

A similar design for a solar roof has been produced by members of the department of architecture at Cambridge University. In November 1973, the group who had produced this design received a £25,000 grant from the Science Research Council, and a further £6,000 from the Department of the Environment to erect an experimental house near Cambridge to investigate the possibilities of solar heating. They claimed that it would be possible to raise a total of 10,000 gallons to an average temperature of fifty degrees centigrade.

Although the solar roofs on both the Dover House and the Washington House were relatively expensive – the materials for each costing about £850 at the time of erection – the general ideas contained in the methods of collecting, storing and distributing heat can be adapted in various simpler ways. One ingenious design, for example, has been produced by a French architect, Jacques Michel. He designed a concrete wall on the side of a house with a large pane of double glass in front of it, and slits at the top and bottom of the wall leading into the house. During the day, air enters the gap between the wall and the glass through the bottom holes, rises as it is heated by the sun, and returns to the house through the top holes. At night the internal surface of the wall radiates into the room the heat that has been stored up in the day during its exposure to the sun. Another idea suggested by Steve Baer in the US is to build a stack of blackened drums containing water into the wall of a house facing the sun. Large hinged shutters are opened to expose the drums to the sun during the day, and are closed at night and during cloudy weather to prevent heat loss through radiation.

Various methods have been proposed by which the sun's energy can be focused to produce high temperatures at a single spot. The general idea is similar to the way in which a magnifying glass can be used to burn a hole in a piece of paper. Experiments have been carried out in Massachusetts, for example, with a solar furnace, in which a large array of mirrors focusing the sun in this way managed to create temperatures of five thousand degrees Centigrade, sufficient to melt a bar of steel. Similarly a solar furnace built at Mont Louis in France has been used to produce an energy source of seventy-five kilowatts. However the technology required to keep all the mirrors aligned in such devices with the sun as it moves across the sky, is considerable. A slightly simpler technique, developed at the University of Arizona, makes use of Fresnel lenses to focus sunlight on to a stainless steel or glass ceramic pipe. Nitrogen is pumped through the pipe to transfer the heat to a central storage unit.

A simple cooking device that uses a large parabolic mirror to focus the sun's rays on to the base of a cooking pot has been developed, primarily for use in tropical countries. Such devices have also been successfully used in sunny areas such as the South of France and the West Coast of the US. It is unlikely that this method would be able to achieve high enough temperatures for cooking in a temperate climate such as the British Isles. The same is probably true of small steam turbines which have been designed to use heat from solar collectors to generate electricity.

These are some of the simpler techniques that have been developed as possible ways of making use of solar energy. Other suggestions that have been made include the generation of heat by means of a heat pump – a type of refrigerator in reverse – using the difference in temperature between the water at the surface of oceans, and that at a great depth where it is considerably colder. There have also been various suggestions for using chemical cells to transform solar radiation into electrical energy. One original suggestion is to make use of the fact that chlorophyll passes electrons to certain semiconductors when exposed to sunlight; it has been estimated that with an efficiency of only ten per cent, chlorophyll solar cells based on this property could produce a kilowatt of energy from ten square metres of surface, at the cost of only a few pence per square centimetre.

Various large-scale proposals for harnessing solar energy have already been worked out in some detail. Two scientists at the University of Arizona, Aden B. Meinel and Marjorie P. Meinel, using the system of Fresnel lenses mentioned above, have calculated that a 1,000 megawatt solar power station, occupying a desert area of over ten square miles, could be built for about just over a thousand million dollars, about four times the cost of a nuclear power plant producing the same amount of power. Even more ambitious is a scheme produced by Peter E. Glaser of Arthur D. Little, Inc., to place a panel of solar cells measuring five miles by five miles in orbit 22,300 miles above the equator. The power would be converted into microwave radiation and beamed back to the earth. For the capital cost of about $500 a kilowatt, roughly twice the cost of a nuclear power plant, it has been calculated that one such satellite would be able to provide for all the energy needs of New York.

At present, the main obstacles to the development of solar energy are not so much technological as economic. Solar energy requires a high initial capital investment, although running costs are relatively low. In addition, schemes such as solar satellites imply the use of relatively large-scale, sophisticated equipment which, although fulfilling the environmental criteria, would almost certainly rule them out of the utopian technology category on other grounds. However, it is always possible to look for alternative ways of using such ideas and adapting them to small-scale use.

WIND-POWER

A second potential source of power is the wind. Over one fortieth of solar energy reaching the earth is converted into wind, which has been used to operate machines for several thousand years. Vertical-axis windmills were in use in Persia many years before the birth of Christ. The Greek inventor, Hero of Alexandria, designed an organ whose bellows were driven by a small windmill, and the Chinese are said to have been using wind for turning their prayer wheels by the fourth or fifth century. The windmill played a central role in the agricultural development of Europe in the Middle Ages, particularly in countries such as Holland whose low topography made the use of water-power impracticable. There are known to have been 10,000 windmills operating in England at the beginning of the nineteenth century, although no more than a handful are still functioning.

Wind-power provides an abundant source of mechanical energy with minimal environmental interference. A major disadvantage is that the speed and direction of the wind are liable to rapid fluctuation, unlike solar or waterpower, with which fluctuations occur over a relatively long period of time. At times, too, the wind may fail completely. However it still remains one of the primary energy sources for utopian technology, and the design of windmills is an area to which an increasing amount of attention is being paid.

There are two ways in which wind-power can be harnessed. The first is the direct use of its mechanical energy. This is the principle on which windmills have been operated for thousands of years to grind corn, as well as for other applications such as powering lathes, saws and other workshop tools, and raising water for irrigation. Windmills can also be used to power heat pumps which extract usable heat from, for example, a passing stream. In general, the use of wind usually involves a windmill connected to a dynamo and additional energy storage equipment incorporating gearing mechanisms and some form of power control, each of which has to be carefully selected to obtain maximum efficiency.

Windmills can be broadly divided into two categories. There are those in which the rotor axis is placed horizontally, the way a propeller is fixed at the front of an aeroplane, and those in which the axis is vertical, like the rotor of a helicopter. The best-known of the horizontal-axis type are the sail windmills that used to cover the countryside of Europe, often standing as much as a hundred feet high. Although their energy output is relatively high, sometimes ten to fifteen horsepower, such windmills are difficult to handle and maintain, and are now little more than relics of the past. A version of the sail-type mill, however, is still marketed by the Danish firm of Lykegaard. It produces up to thirty kilowatts of energy and has a life-time of twenty years or so, although its price of about £2,500 puts it near the extreme end of the utopian technology category.

The commonest type of windmill in current agricultural use is the small multi-blade windmill. Although these do not have a very high energy output, they have the advantage of turning in winds of no more than a few miles an hour. Such windmills are widely used in the Australian outback, for example, to raise water for unattended flocks.

The efficient generation of useful amounts of electricity requires a high-speed windmill to minimize the amount of power lost through gearing. For this purpose 'aerodynamic' windmills with no more than three or four specially-moulded blades are usually employed. It has been calculated that a high-speed windmill eighteen feet in diameter, erected in an area whose average windspeed is twelve miles an hour – as in many coastal regions of the British Isles – could provide more than eighty per cent of the average household's energy consumption for domestic lighting. A typical four-bladed high-speed windmill marketed in the United Kingdom requires a windspeed of sixteen miles an hour for full power, producing an output of 2,000 watts at 110 volts.

A major problem with many high-speed windmills is maintaining control over the speed of rotation. Some achieve this by a feathering effect which reduces the area of blade presented to the wind as the speed of rotation increases. Other windmills have been designed with friction or centrifugally-operated brakes, and in addition many have a fail-safe mechanism by which the windmill falls out of the wind if the windspeed becomes too high. Without such control mechanisms, many aerodynamically-designed windmills have come to a spectacular and untimely end.

The Savonius rotor, developed by a Finnish engineer, S. J. Savonius, between 1925 and 1928, is perhaps the best known and most widely used of the vertical-axis windmills. It is essentially a hollow cylinder which has been cut in half through and parallel to its axis. The two halves are then displaced sideways by about a radius and a half, almost – but not quite – sufficiently to create an S-shaped cross-section, and rotated about a vertical axis through the centre. A fall in pressure on the side away from the wind – known as the Magnus effect – makes the Savonius rotor turn at the same speed as an ordinary rotor with four times its area presented to the wind. The simplicity of the Savonius rotor makes it one of the cheapest forms of windmill. The Brace Experimental Station in Barbados, for example, has designed such a rotor using two forty-five gallon oil drums each divided into two halves. It has estimated the total cost of materials needed to be little more than twenty pounds. A report from the station comments that the Savonius rotor is particularly suitable for irrigation in underdeveloped countries due to its relative simplicity, pointing out that the rotor, drive and pump are eminently suitable for home construction, using only a few tools and a welding set. It claims that the rotor will operate successfully in areas where the windspeed is in the region of eight to twelve miles an hour. Others have suggested the use of the Savonius rotor to power small boats.

One of the main difficulties of harnessing the wind's energy is the problem of storage. Wind is the most variable of all natural energy sources, and some form of energy storage is necessary to guarantee a relatively steady output of power. One commonly-used idea is to put together a number of six or twelve volt car batteries. Although at present a relatively expensive form of energy storage, this technique could benefit from technological advances such as high capacity heat storage materials recently developed in Holland. These materials involve mixtures of fluorides based on common metals such as lithium, sodium and magnesium. They are claimed to have about thirty times the energy capacity of a standard lead-acid battery.

A slightly more sophisticated idea is to use the variable power output of a wind generator to decompose water electrolytically into its chemical components of hydrogen and oxygen. These are stored separately under pressure, and recombined in a fuel cell to provide a steady output of electricity. Research scientists at the Cranfield Institute of Technology are already working on designs for a hydrogen-powered car. Another idea is to use the hydrogen to power a gas turbine engine, as is now done in rockets, which would then turn a conventional generator. Summers comments, 'A wind-driven hydrogen-rocket gas-turbine power plant should be unconventional enough to please the most exotic taste.'[17] It has even been suggested that liquified hydrogen prepared by wind-power in the Aleutian Islands, where the winds are very strong, might be transported by tankers to the Californian coast as an economical form of power!

Many other ideas for harnessing windpower have been put forward. Two Canadian scientists, to take but one example, have developed a high-speed, vertical-axis windmill, described as looking like 'an overgrown barrel-hoop caught sideways on a small radio tower'. The rotor is made of two six-inch blades curved into a 15-foot diameter, and is claimed to produce seventeen kilowatt hours of energy a day with winds averaging about sixteen miles an hour, at the cost of a couple of cents a kilowatt hour.

WATER-POWER

A third source of energy that can be tapped with minimal environmental interference is water. Water-power can be obtained directly from fast-flowing rivers or streams. Alternatively, where the flow is relatively slow, the speed can be increased by building a dam or weir, or by artificially narrowing the width of the channel through which the water is flowing. Like the use of wind, techniques for harnessing water-power have been around for many thousands of years, the basic principles being known and documented by the Greeks.

Water has the advantage over wind in that its flow is far more regular and predictable. It was largely this that made it the first major source of power to be used for industrial applications, particularly during the Middle Ages. A disadvantage of water-power is that without means of transmitting power over long distances, its use is restricted to the locality around rivers and streams, giving none of the geographical flexibility associated with wind-power. Furthermore it obviously cannot be used once the temperature falls below freezing-point.

As with wind, water-power can be utilized either directly or indirectly. The direct use is usually achieved with a waterwheel, the major source of water-power up to the middle of the last century when it was superseded by the hydroturbine. There are three types of waterwheel: the over-shot wheel, in which the water is fed, by means of a narrow channel known as the mill-race, over the top of the wheel; the under-shot wheel, in which the water passes under the wheel; and the breast-wheel, which was very common in Northern England in the early nineteenth century, and in which the water meets the wheel roughly at the level of the wheel's axis, subsequently travelling vertically downwards and out under the wheel. Waterwheels can be very powerful. As early as the second century AD a vast complex at Arles in France comprising sixteen waterwheels is said to have been capable of grinding thirty tons of flour a day. However the relatively slow speeds of most waterwheels, and the elaborate gearing that this often makes necessary, present obstacles to their use as a viable source of energy.

An alternative way of harnessing waterpower is through the use of turbines. Unlike the waterwheel, which is primarily turned by the weight of the water, the turbine is designed with special vanes that effect the greatest possible change in the direction of motion of the water. The turbine derives its energy from the impulse of the water against the vanes and the reaction that results – according to Newton's laws of motion – from its change in direction. The high rotation speeds make the hydroturbine eminently suitable for the generation of electricity, and hydroelectricity is widely produced in many countries. Large hydroelectric schemes often cause widescale environmental damage, particularly through the creation of vast artificial lakes required to provide a sufficient 'head' of water. Small-scale water-turbines however can be installed at the side of a river with minimum interference to the overall flow. The firm of James Leffel & Co. in Springfield, Ohio, for example, has been producing such small-scale power-generating turbines for over a hundred years. One advantage of water-turbines over waterwheels is that the electric power generated can be transmitted relatively easily to wherever it is needed (although this obviously raises the level of the technology required).

One use of water to produce energy in coastal areas is by harnessing the power of the tides. Water is allowed to flow into a reservoir at high tide, and out again through a waterwheel or turbine once the tide has receded. Tide-mills have been separately developed both in Europe and the US. During the eighteenth century, for example, a tide-mill in Virginia used the tides of East River to grind meal and flour. In the 1930s a scheme was put forward to make use of the tides in the Bay of Fundy, between Maine and Canada, which it was claimed would provide about 300 megawatts of power. The French government has recently built a tidal powerplant in the estuary of the Rance River, whose tides average twenty-seven feet, that produces an output of 240 megawatts of power.

It is also possible to use the rise and fall of waves as a source of energy. A wave-driven motor, referred to as the Sea Horse, has already been built capable of operating a twelve volt car generator. A pair of simple pistons, operating where waves were eight feet in height, has been found to provide about 700 watts of usable energy. By applying the same principle on a much larger scale, it has been suggested that it might be possible to build a hundred-piston power-generating station at sea which would produce 3,700 kilowatts.[18]

METHANE GAS

A further ecologically-sound source of energy which, unlike either wind or water, has been little used in the past, is methane gas. This is produced by the decomposition of organic waste such as farm manure, kitchen waste or human excrement. The method used of obtaining the gas is basically an extension of the technique used for the digestion of sludge in municipal sewage tanks. The waste material is usually placed in a tank sealed from atmospheric oxygen, and heated gently. After about a week it begins to give off methane, which is collected in some form of storage holder. The gas can be used as a fuel for cooking, lighting, heating and refrigeration, and even for powering small engines. It has been calculated that every pound of waste treated should produce between one and four cubic feet of gas, providing about half a kilowatt hour of energy. Various designs have been produced for inserting methane-producing tanks into home sewage systems as part of a general waste recycling process. The manager of a sewage works in Cheshire already runs the central heating in his home on the waste methane gas which would normally be burned in the works chimney. A Devon farmer, Harold Bate, even markets a small converter which enables an ordinary car to be run on a mixture of petrol and methane. The methane is stored in a cylinder in the car's boot, and results in greatly reduced levels of polluting emissions, although also a slight reduction in the car's power. Most methane gas generators have a relatively low energy output, although a group of research scientists in Scotland is now working on a design which, it is claimed, will eventually be able to produce about 300,000 BTUs a day out of the waste from 100 pigs[91].

These four sources of energy – solar energy, wind-power, water-power and methane gas – together with the use of animals and human labour, would, it is suggested, provide the power base of utopian technology. The total amount of energy available from these sources is likely to be considerably less than that at present consumed by advanced industrial societies. But the amount of power required by utopian technologies is itself likely to be equally reduced. In other words, rather than looking for alternative ways of producing the same amount of energy that we use at present or are predicted to require in the future, it is possibly more appropriate to develop technologies whose energy requirements are compatible with the amount that would be available from these alternative sources.

Used in conjunction with each other, the sources described above could probably meet all the energy needs of small communities. Besides providing a sound basis for harmonizing man's technological activities with environmental requirements, they might also bring a more natural rhythm back to these activities themselves. As Murray Bookchin has written, 'to bring the sun, the wind, the earth, indeed the world of life, back into technology, into the means of human survival, would be a revolutionary renewal of man's ties with nature.'[20]

5 Utopian Technology:
commodity production and social organization

FOOD

The production of food is one of the most important factors affecting the ecological balance. A major consequence of the Industrial Revolution, with its rapid growth of urban populations, has been the accompanying development of intensive food-production techniques. These have turned agriculture in many countries from a way of life into a technologically-based industry. Modern agriculture is characterized by the massive use of artificial fertilizers to promote the growth of crops; the increasing use of factory-farming methods of raising animals in totally artificial conditions; a dependence on single crops – known as monoculture – resulting in greater economic efficiency than could be achieved with traditional patterns of crop variety, but often leading to the destabilization of important parts of the ecological cycle; and the general mechanization of all aspects of agriculture, further widening the gap between man and nature.

Utopian technologists seek to reverse these trends. They are concerned to develop an agriculture that adequately meets the needs of the community, yet at the same time enables man and the natural environment to exist in harmony, rather than opposition. Great emphasis, for example, is placed on the need to return to systems of organic farming, by which animal and other organic waste is used to fertilize the soil, thus resulting in the minimum waste of valuable nutrients and avoiding the use of artificial chemical fertilizers. Ways in which cultural variety of crops can be arranged to fit in with natural ecological cycles are being actively investigated. So, too, are various farming techniques which do not require the use of sophisticated equipment, yet still provide agriculture with a sound and efficient technical base. Finally, the great potential offered by small-scale market-gardening is heavily stressed. Back-gardens and allotments are still claimed to have the highest productive yield per acre of all agricultural land. This is largely due to crop rotation and personal involvement as opposed to the monoculture and mechanization practised in conventional agriculture. The whole utopian approach to farming was aptly summed up by the eighteenth-century French utopian, Charles Fourier. 'It is impossible to organize a regular and well-balanced association,' he writes, 'without bringing into play the labours of the field, or at least gardens, orchards, flocks and herds, poultry yards, and a great variety of species, animal and vegetable.'[1] Murray Bookchin suggests that 'agriculture will become a living part of human society, a source of pleasant physical activity and, by virtue of its ecological demands, an intellectual scientific and artistic challenge.'[2]

Traditional farming techniques have much to teach the utopian technologist. The value of skills and knowledge built up over thousands of years, and only lost over the past century or so, is slowly being acknowledged. At the same time research is being carried out into ways in which existing scientific and technical knowledge can be used to intensify agricultural productivity in an ecologically sound manner. A typical example of this is a system to integrate the raising of fish and the growing of vegetables, designed for a small co-operative community in the New Mexican Rockies by the New Alchemy Institute in the United States. This system involved the use of two ponds, dug side by side. One pond is heated by solar energy, and stocked with *Tilapia*, a good eating fish which feeds on photoplankton grown on the pond. The pond is also designed to contain crayfish and catfish, with ducks on the surface and mussels on the pond's bed. The mussels provide considerable amounts of food and also filter the water in the pond, while their waste products constitute a powerful fertilizer. Thus a system of rotation is planned by which on alternate years the second pond is used for the intensive raising of vegetables, and the functions of the two ponds reversed annually.

The New Alchemy Institute has also studied methods of indoor fish-farming using ponds covered by plastic domes. Each of these domes has a double skin to improve insulation, as well as a venting

system to make temperature control possible. Again the fish used
are *Tilapia*, and a number of experiments have been carried o
into different types of food. The most successful source of food f
the fish was again found to be planktonic algae produced in t
pond, but they also showed a liking for carrot tops and grour
soya beans. The latter are commonly fed to fish under culture
the Orient, and contain extremely high levels of protein. (Simil
plans for a cat-fish farm have been drawn up by, among other
an American artist Newton Harrison. Six tanks, some of whic
also included shrimps and crayfish, were exhibited as a 'lif
support system' in an exhibition at the Hayward Gallery i
London in the autumn of 1971. The total exhibit included a me
at which the fish were to be fried and eaten. Newton Harriso
caused considerable consternation in the British press when l
revealed that the fish were to be electrocuted in a speciall
designed box, despite his assertion that this was the most human
way of killing them).

Plants can also be grown indoors. One technique is known a
hydroponics, by which plants are grown, not in soil, but in wate
to which suitable quantities of nutrients have been added. Th
technique has the advantage that the roots of the plants, bein
immersed in liquid, are always in contact with a considerabl
amount of nutrient. The nutrients required by most plants includ
balanced amounts of nitrogen, potassium, phosphorous, calciun
magnesium and sulphur. All of these are usually present in th
soil; they can be used hydroponically in the form of soluble salt
such as potassium phosphate and calcium nitrate. Much smalle
amounts of elements such as iron, manganese, boron, zinc an
copper are also required, these being known as trace elements.

A major disadvantage of growing plants in solution is the diff
culty of ensuring an adequate supply of oxygen to the roots, an
also in getting the plants to stand up straight. One solution is t
grow the plants upright in an aggregate such as sand, to which th
nutrient solution is then added. High plant yields can be pro
duced relatively easily in cities, or areas where the soil is poor, b
the use of hydroponic techniques. In particular, the technique
could have wide applications in underdeveloped countries whic
suffer from a lack of fertile ground. However, great care has to b
exercised in maintaining the right balance of chemical nutrient

It can be argued that the artificiality of hydroponic techniques makes them open to similar criticisms to those aimed at factory-farming methods of rearing animals, and the food produced is claimed by some to be considerably less wholesome than that grown naturally.

The traditional way of increasing agricultural yields is through the use of fertilizers. A valuable form of fertilizer can be produced from sewage and other organic waste, processed to form agricultural compost. We have already described a method by which it is possible to extract methane gas from such waste. An important part of this process is that the humus that remains after the methane has been extracted can be dried out and used as a powerful fertilizer, claimed to be at least four times richer in nitrogen and phosphorus than most artificial fertilizers. A Swedish designer has produced a composter known as the Clivus which consumes both sewage and kitchen waste, and produces a rich, dry compost which can be spread straight on to the land. This composter is claimed not only to be cheaper than most septic tank installations, but also to reduce the water requirements of the average household by twenty gallons a day.

A more direct way of making agricultural use of sewage has been developed by an interdisciplinary team at Pennsylvania State University in the US. This involves extracting the suspended solids from the sewage and decomposing the compounds that are left in solution. The waste sewage water is then sprinkled on forested and agricultural areas. The nutrients from the water are absorbed into the plants, while the water itself is purified as it drains through the soil, and can be drawn up through wells in a drinkable condition. Experiments conducted by the research group have revealed that the yield of corn and oats can be increased by between fifty and a hundred per cent by spraying with two inches of waste sewage water a week.

Many other ideas have been put forward for alternative methods of food production. One is the concept of 'three-dimensional agriculture' by which a variety of plants growing at different levels above and below the ground is used to intensify the output of a given area of land. Many tribal societies, for example, grow a complex selection of plants chosen in this way. Using such a system, deep roots, shallow roots, ground-runners, short stemmed plants and long-stemmed plants can be grown almost on top of each other. The systems practised by tribal societies have usually been developed over many centuries, and result in the maintenance of both soil fertility and genetic variety in the plants with the minimum amount of labour.

A second idea is the use of soya beans as a source of protein. The soya plant takes up little land, and it costs only a fraction of the price of the same protein yields of animal products. Also, since it obtains its nitrogen from nitrogen-fixing bacteria in its root nodules, it does not require the use of artificial fertilizers. The soya plant is usually grown in the tropics, but can also be successfully grown in temperate climates. Other ideas for alternative methods of food production include the extraction of protein from green leaves, and its artificial synthesis from crude oil, although the latter seems to require a relatively sophisticated set of techniques that might place it outside the utopian technology category.

SHELTER

There are a number of alternatives to conventional forms of building. The main considerations here are to avoid, wherever possible, the use of scarce or non-renewable resources, and to build in a way that does not require the use of sophisticated engineering techniques. There are advantages, too, of building with materials that are locally available. Most of these are traditional building materials, such as wood and dry stone. Earth can also be used for building, especially if it is rammed tightly into place or 'stabilized' by the addition of a little cement.

Mud and vegetation are widely used in tropical countries, although these are unlikely to make adequate building materials in colder, wetter climates without additional treatment. A fascinating book entitled *Architecture without Architects*, by Bernard Rudofsky, illustrates houses made out of bamboo poles and thatch, giant reeds and woven matting. It provides examples of 'vernacular' architecture from all parts of the world, ranging from underground houses excavated out of porous rock in China to the use of giant, hollowed-out baobab trees, up to thirty feet in diameter, in Africa.

Utopian technologists see little need to stick to conventional

niques – have been widely discussed. Hamza Alavi, for example, writes that 'every effort is made to influence the bureaucracy (of the underdeveloped countries) in favour of policies which are in conformity with metropolitan interests. This is ideologically expressed in the form of "techniques of planning" and it is presented as an objective science of economic development. The western-educated bureaucrat is regarded as the bearer of western rationality and technology: and his role is contrasted with that of "demagogic" politicians who voice "parochial" demands'.[14]

To take an example, which demonstrates the extremes to which these techniques can be logically pressed, we can refer to a paper entitled 'Economic Worth of Preventing Death at Different Ages in Developing Countries', by two members of the US Cosmos Club, Dr Stephen Enke and Mr Richard A. Brown. The authors argue the importance of rationing scarce medical resources in the underdeveloped countries. They point out that every living person has a particular capitalized value to the economy, based on his expected additions to national output and subtractions from it as a consumer. 'Why should publicly financed resources be devoted to preventing infant mortality,' they ask, 'when the economic worth of such marginal infants is negative? The economy would be better off without them. The burden of proof is surely on those who recommend division of health resources from caring for producing adults to caring for consuming children.'[15] When we see this type of 'scientific' argument applied to justify the stepping up of birth control campaigns, we begin to recognize the repressive nature of the ideological interpretation on which such forms of analysis are frequently based. The task should be to produce a social system that meets the demands of people, not people that meet the demands of a system.

In many ways, both utopian and intermediate technology provide an antithesis to the systems analysis approach. They refuse to accept that social life can be fragmented into isolated parts. Schumacher himself has even argued the case for a 'Buddhist economics'. This would be based on a non-materialist concept of the importance of work to both the individual and the community, and embrace 'the systematic study of how to gain given ends with the minimum means'.

But despite such attempts to provide an alternative ideology for intermediate technology, in practice it finds itself all too easily swept up into the prevailing ideology, as a direct extension of the concepts of industrialization discussed previously. ITDG reveals this continuity with dominant ideology when it declares that 'we are now beginning to realize that development is an organic growth and that *to intermediate levels of development there must correspond intermediate levels of technology*'.[16] The use of the term 'organic growth' implies a neutral development process independent of political considerations concerning the distribution of power and control between either classes or nations, an attitude which we have previously characterized as the ideology of industrialization.

Part of this ideology is the view that the relationship between the industrialized and the underdeveloped countries can be seen as one of mutual co-operation, whereas in practice it is only too often one of political and economic domination. When, for example, promoters of intermediate technology declare that the choice of technology is one of the most important choices that a developing country must face, one must ask: choice by whom, and for whom? The economic situation in most underdeveloped countries is determined by the united economic and political interests of foreign capital and an indigenous élite. The rural poor, as non-surplus producers and non-consumers within an emergent capitalist system, become increasingly irrelevant to the political process. Any claim for the democratic choice of technology in such a situation has a very hollow ring about it indeed. What is often meant is that the choice is one that faces foreign aid and investment bodies, and the concepts and ideologies that support them; it is a political as much as a technological question, and to view it solely as the latter is but one further example of ideological distortion.

ITDG states similarly that 'the group's special emphasis is on the commercial viability of all projects . . . the experience and know-how of established firms in industry and commerce is being mobilized to this end'. In other words, intermediate technology must coincide with the existing economic system; in Marxist terms the problems of development are to be solved merely by transforming the modes of production, leaving the relations of production intact. Attempts have in fact been made to fit intermediate technology into a framework of neo-classical economics.

Economists have tried to demonstrate that intermediate technology makes good economic sense in terms of the traditional production function. Others have pointed out their value in helping to develop a capitalist economy through the training given to entrepreneurs. According to K. L. Nanjappa, for example, Development Commissioner for Small Scale Industries in India, 'Apart from its vast potentiality to absorb the rapidly increasing labour force, these industries have also encouraged the competitive element . . . the growth of an entrepreneurial class requires an environment and small-scale industries provided that environment by encouraging the creation of a network of feeder and complementary relations among different enterprises. It is in this environment that latent talent of entrepreneurs provided self-expression in localized innovations and cost-saving measures.' An example of this is provided by the ITDG project on building in Nigeria which concentrates on training small-scale contractors in business management techniques.[17] Intermediate technology can quickly become the seed-bed for small-scale capitalism.

Looked upon as an extension of the ideology of industrialization, it becomes possible to challenge two important empirical assumptions on which the concept of intermediate technology, and in particular its relevance to the problems of underdevelopment, are based. The first assumption is the implied relationship between the introduction of capital-intensive production techniques into the industrialized sector and the rise of unemployment. According to Professor Hans Singer, for example, 'if you put imported capital-intensive technology together with the high birth-rate you have an explosive situation, because they simply do not fit together'.[18] This and similar statements imply a direct causal relationship between capital-intensive patterns of industrialization and the rise of unemployment.

Yet such an explanation is insufficient. It is important to point out, for example, that the relatively low price of labour in many developing countries can make what appears to be capital-intensive machinery in the industrialized countries relatively less so when the same machinery is incorporated into the economic environment of underdeveloped countries. A study of silk manufacture carried out by Gustav Ranis of the Economic Growth Center at Yale University found that techniques requiring one girl to operate up to seven looms in Japan are applied in South Korea where the wages are much lower with one girl operating only two looms.

If we are to seek the true causes of unemployment and under-employment, we must go further than merely laying the blame on the introduction of capital-intensive technology. It has been pointed out, for example, that usually indirect evidence, such as wage and capital costs and competitive assumptions are used to justify estimation procedures leading to the conclusion that capital 'deepening' is the major cause of the slow growth of employment in the underdeveloped countries. 'While this interpretation may be correct,' suggests one economist, Howard Pack, 'it requires a strong leap of faith given the lack of any adequate micro-economic evidence and the absence of strong competitive pressures in most less developed countries.'[19]

Studies of industrial techniques in Kenya have led Pack to the tentative conclusion that even within the modern sector, Kenyan production appears to be relatively labour-intensive. He suggests that alternative factors, such as under-utilized plant, may have a considerable short-run effect on urban unemployment. His conclusions illustrate that unemployment is a complex social and economic problem: it is unlikely to be solved merely through the implementation of industrialization policies based on concepts of appropriate or intermediate technology, but ignoring their social and political dimension.

A second empirical assumption of intermediate technology is the concept of the dual economy. This implies that the economy of a developing country can be treated as if composed of two, independent parts, the modern and the traditional (or non-modern) sector. Supporters of intermediate technology stress that they are primarily concerned with the traditional sector, usually, but not necessarily, identified with rural rather than urban areas. They see intermediate technology as a complementary strategy to the advanced technology being introduced into the modern sector. 'If there is a dual economy, why not have a dual development strategy?' asks Mabdul ul Haq, a former chief economist with the National Planning Commission in Pakistan, and now an adviser to the International Bank for Reconstruction and Development. The assumed existence of a dual economy can be challenged on both empirical and theoretical grounds. A number of authors have suggested that, in a historical sense, the concept of rural development may in fact be a direct function of the development of towns, and that rural economies have been built up in response to the needs of towns for food and other commodities, rather than towns resulting from the production of agricultural surplus.

Even though it may be possible to identify separately the modern and the traditional sectors of an underdeveloped country's economy, this by no means implies an independence between them. Indeed the domination of the traditional by the modern sector is often an important structural element in the economy itself. The economist, Andre Gunder Frank, points out that the industrialized countries have always suppressed indigenous technology in the underdeveloped countries which conflicted with their interests. He gives as examples the way in which Europeans destroyed the irrigation and other aspects of agricultural technology in India, the Middle East and Latin America, and the way the English suppressed indigenous industrial technology in India, Spain and Portugal. 'The same is true on the national and local levels,' he writes, 'in which the domestic metropolis promotes the technology in its hinterland that serves its export interests and suppresses the pre-existing individual or communal agriculture and artisan technology that interferes with the use of the countryside's productive and buying capacity and capital for metropolitan development.'[20] It is the patterns of world trade, largely dominated by the interests of a foreign capitalist class, that determine both the external and the internal structures of the economies of most underdeveloped countries. Within this framework, it is theoretically impossible to talk of the independence of the modern and the traditional sectors of the economy. In real terms, the situation is as likely to be one of conflicting political demands such as centralization against regionalization. These have often dramatic consequences – witness the recent independence struggles of Bangladesh.

It is not difficult to show how the patterns of world trade are directly reflected in the mechanisms which determine the process of technological choice. The most important of these mechanisms is the concept of technology transfer, by which technology is transferred either directly or indirectly from the industrialized to the underdeveloped countries. Many countries pay more for patents, licences, and technical know-how from the industrialized countries than they do on their own research and development programmes. Ceylon, to take a typical example, spent 0.5 per cent of its gross national product in 1970 on such forms of technology transfer, but only 0.2 per cent of its GNP on its own R & D programmes; the same pattern is true for many other underdeveloped countries. A recent report from the UN Conference on Trade and Development (UNCTAD) estimated that the underdeveloped countries in general spend on average a sum equivalent to five per cent a year of the total value of their exports on patents and licences; if current trends continue, UNCTAD has calculated that this could rise to fifteen per cent of their exports by the end of the 1970s.

Technology can be transferred in two ways, either directly to a firm in the underdeveloped country, or indirectly through the subsidiaries of multinational companies. In both cases, the nature of the technology transferred is almost entirely determined by those whose interests are unlikely to coincide with those of the majority of the inhabitants of the underdeveloped country concerned. Multinational companies often dominate the key industrial areas in a developing country, and hence the process of technological innovation in these areas is primarily oriented towards maintaining the profits of major share-holders, who generally live outside the country. The situation is little better in the case of indirect transfers, since the patents and licences covering technological innovation often carry heavy restrictive clauses. A study of thirty-five contracts made by firms in Bolivia, for example, for the transfer of technology from firms in the industrialized countries, found that twenty-four tied technical assistance to the use of patents, twenty-two tied additional know-how to the existing contract, three fixed the price of final goods, nineteen required secrecy on know-how during the contract and sixteen after the end of the contract. As the Junta del Acuerdo de Cartagena, who carried out a study on policies relating to the transfer of technology in South America for UNCTAD, pointed out in their report, 'the only basic decision left to the licensee is whether or not to enter into agreement for the purchase of technology. Technology, through the present process of its commercialization, becomes thus a mechanism for controlling the recipient form. *Such control supersedes, complements or replaces that which results from ownership of the firm's capital*' (my italics).[21]

The message of these and similar facts is clear. Technology has now become an important mechanism by which united capitalist interests maintain their dominant economic control over the poor countries. We found previously that the supposedly objective process of technological innovation in the industrialized countries was largely determined by the need to maintain social control over the work-force. The process of technological choice performs a similar function in maintaining control over the activities of the underdeveloped countries, again securing their maximum exploitation in the interests of foreign and national élites. A Brazilian sociologist, Fernando Henrique Cardoso, for example, using the argument that imperialist economies need external expansion for the realization of their capital accumulation, suggests that within the dependent economies, capital returns to the metropole in order to complete the cycles of capitalist reproduction. 'That is the reason why "technology" is so important,' he explains. 'Its "material" aspect is less impressive than its significance as a form of maintenance of control and as a necessary step in the process of capital accumulation.'[22]

Technology transfer is more than just an articulation of the economic relationships between the industrialized and the underdeveloped countries. It is also an important means by which these relationships are maintained and controlled through the constraints imposed on technological choice. The neutral disguise is stripped away to reveal the political motivations at work, and once again we see the identity of technology as providing the political mechanism of social control. 'Far from diffusing more and more important technology to the underdeveloped countries,' writes Andre Gunder Frank, 'the most significant technological trend of our day is the increasing degree to which new technology serves as the basis of the capitalist metropolis' monopoly control over its underdeveloped economic colonies.'[23]

The ideology of industrialization becomes absorbed in the cultural patterns of those underdeveloped countries on which it is imposed. An interesting illustration of this is provided by a study of technological choice carried out on a sample of fifty industrial plants in Indonesia. These plants covered industries such as cigarettes, soft-drink bottling, plastic sandals and bicycle tyres. The study, carried out by L. T. Wells, found that often the technology selected in any one instance was *not* the one most appropriate to the prevailing economic aspects of the relevant productive process. On the contrary, he found that, for example, one important factor was the use of international trade names. These often overshadowed the price advantages of competitive intermediate and labour-intensive technologies. Perhaps most important, however, was an apparent built-in bias of managers and engineers in favour of capital-intensive advanced technology. When a firm had a monopolistic advantage in local markets, according to Wells, the 'engineering man' seemed to be allowed to over-ride the 'economic man', and he comments that 'the catalogue issued by the European or United States exporter of machinery is still the prime source of technical assistance' in many underdeveloped countries.[24]

A further example of the extent to which the ideology of industrialization, which is equated with both development and modernization, has penetrated into the culture of underdeveloped nations is provided by the results of a survey investigating the ambitions of schoolchildren in the rural areas of the West African state of Togo. The majority of those interviewed replied that they would like to be civil servants, and that they would like to have a car, if possible a Mercedes. According to one twelve-year-old, 'I would like to be a mechanic so as to earn a good living. I shall live in Lome (the capital of Togo) in order to have a large number of cars to repair; I'll earn a lot.' Such is the stuff that dreams are made of in many underdeveloped nations.

As in previous chapters, the main conclusion that we must draw is that, to the extent that technology becomes, not only a function of, but in fact an integral part of the dominant political interests, it is impossible to think in terms of technological change unless we are simultaneously prepared to consider the need for social and political change. Indeed, it is only through political change, and in particular through achieving liberation from the economic and political shackles of a dominant class, that the possibility of significant technological change can emerge. The category of 'inappropriateness' in technology, as that of underdevelopment in general, is the inevitable function of a world trade system that reinforces the supremacy of the united interests of foreign capital and an indigenous élite: technology becomes one of the prime means by which this supremacy is maintained.

Within this perspective, we see clearly why intermediate technology, although in principle admirably suited to the conditions prevailing in the traditional sector of the economy of many underdeveloped countries, is unlikely to be successfully applied in practice unless accompanied by the necessary structural political changes. These changes are not merely of a cosmetic and superficial nature, but reach to the heart of the political systems of the underdeveloped countries. They therefore relate as much to changes in the modern sector as to those demanded by the traditional sector, if a situation in which the two sectors work in co-operation is to be substituted for the present one in which the one dominates and suppresses the other.

To take a specific example, supporters of intermediate technology often point to China as an example of a relatively underdeveloped country in which a pattern of dual development embracing both modern and traditional sectors of the economy has been achieved. There are many instances of ways in which the Chinese have been able to use human labour to carry out tasks that would have been mechanized in the West, and Chinese technology is frequently referred to as a prototype of genuine intermediate technology. It is important to realize, however, that this has only been achieved since the People's Revolution in 1948, and in particular since the Great Leap Forward that began at the end of the 1950s, during which the Chinese broke with their Soviet technical advisers and Mao Tse Tung first put forward his 'walking on two legs' strategy for industrial and technological development. This strategy refers to the technique of combining agricultural with industrial development, new with traditional production techniques, and small-scale labour-intensive local industry with large-scale capital-intensive modern industry. Much of China's advanced technology was originally imported from the Soviet Union, Japan and Europe, and it has been estimated that, next to the US and the USSR, China must soon be considered as the third most important technological power in the world.

How relevant is China's experience to the situation of other underdeveloped countries, faced with a similar set of initial conditions? It is important to stress that science and technology in China cannot be examined apart from the social, economic, political and ideological setting in which they are pursued. It has been pointed out that Chinese planners have set about solving the type of problems now faced by many developing countries by imposing social controls which policy-makers in other societies may not have access to. These range from controls on patterns of consumption and the level of real wages to the redistribution of income on egalitarian grounds. They are closely tied to the structure of political power and the social objectives of those responsible for policy-making. It appears to be the non-competitive structure of the Chinese economy which enables the sharing of technology and technological know-how between enterprises, and the protection of the markets of small-scale industries.[25]

Other countries in which an intermediate technology has already become an important part of official development strategy are Cuba and North Vietnam. Achievements of North Vietnamese scientists and technologists, for example, working under war conditions and on severely limited resources, range from the development of an efficient transport system based mainly on bicycles, to the breeding of various water plants for use in flooded fields as an alternative to chemical methods of nitrogen fixation.[26] The distinctive feature of these countries is that they have created, often only after bitter struggle, an economic system based on a planned, rather than a market, economy. This would appear to substantiate the suggestion that any alternative to the advanced technology that characterizes the industrialized countries, presents a political as much as a technological challenge to the existing economic system.

When it becomes identified separately from the need for political change, intermediate technology become little more than a vehicle of economic and cultural imperialism. Through the mediation of aid and investment bodies, it permits the industrialized countries to continue to impose not only their cultural values and ideologies but even, as we have seen in the case of technology transfer, direct economic control over the underdeveloped countries. They are often supported in this by the political control held by small Western-oriented élites within the underdeveloped countries themselves. The educationist Ivan Illich has warned that 'the cultural revolutionary must . . . be distinguished from the promoter of intermediary technology who is frequently merely a superior tactician paving the road to totally manipulated consumption'.[27]

An example of the way intermediate technology can be used to further neo-colonialist interests is the development by the Ford Motor Company of a vehicle specifically designed to meet the environmental and economic conditions of South-East Asia. The design brief for the vehicle specified that it should be of low cost, and that it could be produced with local materials, simple equipment and local labour – in fact the general criteria associated with intermediate technology. The result has been a small pick-up truck called the Fiera, claimed to be suitable for many applications in farming and light industry. According to William O. Bourke, president of Ford Asia-Pacific Inc., 'in a very real sense, Fiera will be Asia's own vehicle, largely designed and wholly built in the Asia Pacific region . . . we believe it will bring the cost of vehicle ownership within reach of many thousands of potential customers in the emerging middle classes.'[28] He quotes with approval the statement of Henry Ford II that 'a corporation can serve society only if it is profitable. And it can stay profitable only if it is responsive to the needs and values of the society in which it operates'. Little mention here, of course, of how Ford intends to respond to the needs and values of the great majority of the populations of underdeveloped countries, those who are not, and never likely to be, included in these emerging middle classes.

Many technological innovations which have been specifically developed to suit the environmental and economic conditions of the Third World, and which have not been accompanied by the necessary social and political changes, have often themselves led to major problems. A prime example of this is the Green Revolution, brought about by the introduction into many underdeveloped countries of highly-productive strains of wheat and rice. Although these varieties have been generally successful in raising overall levels of food production, the benefits resulting from this have been far from evenly distributed. Even Robert Macnamara has admitted that 'the social and economic challenge is to prevent the benefits of the Green Revolution from being monopolized only by the wealthier farmers. So far, the more advantaged farmers have obtained disproportionate shares of irrigation water, fertilizers, seed and credit. Unwise financial policies have sometimes encouraged these farmers to carry out excessive mechanization'.[29] In 1971 the growing problems associated with the Green Revolution forced the UN Development Programme and the UN Research Institute for Social Development to launch an intensive, integrated study of the social and economic impact of the new cereal strains.

These are the type of problems that can result from an inadequate analysis of the roots of a particular social issue. They often result from attempts to impose a Western-oriented 'scientific' assessment to problems that involve their fragmentation from a general social and cultural environment. Numerous examples can be found of attempts by aid agencies to tackle problems associated with public health and nutrition, for example, where it has been only too tempting to treat the immediate symptoms rather than the root causes.

In general attempts to solve urgent social problems through the direct application of what are considered to be appropriate technological solutions have met with, at best, only limited success, and at worst, a general deterioration of the situation. Unless complementary political changes are introduced at the same time, technological solutions, however appropriate, are unlikely to provide adequate solutions by themselves. Political changes will neither flow automatically from, nor be determined by, the technology. They must be introduced separately as part of the general political struggle for emancipation. Truly appropriate technology can only come from the demands of the people by whom and for whom it is to be used, once they have successfully realised their own political and economic strength.

Here we confront the dilemma of intermediate technology. Both from the analysis we have sketched out, and from events such as the failure of the UNCTAD conference on world trade held in Santiago in April 1972, we can see that there is little hope of the underdeveloped countries deriving much profit from participation in the international scene. Nor can the industrialized countries be expected to pursue voluntarily policies that do not further their own economic or political interests. The alternative for the under-developed countries is to adopt development strategies that concentrate on concepts of self-help and, as far as possible, the use of locally available resources. In doing so, the industrialized countries could be forced to rely less heavily on the import of raw materials from the underdeveloped world, and hence obliged to develop their own self-reliance policies along the lines of the utopian technology described in the previous two chapters.

Intermediate technology is ideally suited to policies that stress the importance of self-help, being based on the general concept of 'development from below'. Yet the implementation of these policies, and the social structures that they imply, is a political task. Intermediate technology can thus only be realistically seen as part of a political strategy aimed at liberating the population of the underdeveloped countries from political domination and economic exploitation, whether coming from the industrialized nations, or from their own, often Western-oriented, élites and bureaucracies. Unless the unity of technological and political practice is achieved, it is unlikely that intermediate technology, on its own, will be able to tackle the real causes of underdevelopment, nor bring about a viable solution.

7 Myths and Responsibilities

Finally we come back to our own situation. Having discussed the historical and utopian aspects of technology, as well as those which concern the underdeveloped countries, it still remains to discuss the implications for those of us who live in industrialized capitalist societies. To do so we need to look closely again at both the material and the ideological role played by technology in these societies. And to understand this, we must also study the way technology is related, conceptually as well as materially, to the practice of science. The issues raised by this discussion lead on to a set of proposals of the type of activities in which it is necessary to engage if we wish to change this situation.

The first section of this chapter summarizes how we might approach the political nature of technology as a social institution. There are two aspects of this. The first, from what might be called the synchronic perspective, attempts to look at a society's technology as it exists at any one point in time. Here I wish to suggest that technology can be viewed as a 'language' of social action, from which we extract those individual elements – whether tools or machines – required to carry out a particular task. A corollary of this is that technology shares the same structure as that of the dominant modes of social action and interaction, themselves reflections of the distribution of power and the practice of social control. The second aspect is taken from a diachronic perspective, and examines the historical development of technology. Here, as before, I suggest the crucial importance of understanding the political nature of the process of technological innovation.

The second section of this chapter deals with the way that society *interprets* the nature of technology, and of technological development. I wish to suggest that this interpretation exists in the form of a myth. A myth provides an apparent historical interpretation of a particular state of affairs – the Romulus and Remus myth, for example, provided an explanation of the origins of ancient Rome. In the case of technology, I suggest that this explanatory function is achieved by the myth whose message is that 'technology is politically neutral'. Furthermore, I suggest that the interpretation this myth implies of both the nature of technology and the process of technological development, is largely derived from a positivist interpretation of the practice of science, and the extension of this into the world of everyday experience, where it becomes expressed as the social philosophy of 'scientism'.

In the third section, I shall indicate how the existence of this myth determines the way we are led to interpret the nature of technology, and some of its implications. In particular, I suggest that technology functions symbolically within a number of what I call 'modes of significance', the various perspectives or frameworks, be they economic, political, artistic or whatever, we use to interpret events or experiences and hence to construct our own social reality. Here, again, I suggest that technology is generally interpreted to function in a neutral manner, merely providing the symbols by which particular notions are expressed or 'signified'. As with the material function of technology, I shall argue that this interpretation disguises not only the essential role that technology plays within each of these modes of significance, but also the fact that to divide experience into such modes is already to distort it, a distortion linked to the atomistic and reductionist tendencies of scientism.

Finally, I have added a brief discussion of how it might be possible to break out of our present situation, to strip away the mythological disguise that surrounds technology and reveal the political reality that lies beneath. This, I argue, can only be achieved through activities that directly challenge the content of the myth of the political neutrality of technology, as well as the power structure which this myth maintains, and on which it relies.

Although technology, as a social institution, has been described as an abstract concept, it is through the action of individual machines and tools that technology is directly experienced.

Furthermore, each machine that we experience is used by an individual or a group as a means of carrying out a particular action in the social domain. If we see a car in the street, for example, we know that it is being used by its driver to make a particular journey in a particular manner; if we see a telephone in a house, we know that this is one way its owner communicates with other members of society; if we see a computer on a factory floor, we know that this is one way management has decided to control a production process. We might say, therefore, that each machine that exists in a society 'objectifies' a particular mode of social action, in other words gives this action an 'objectified' form. A car can be looked at as the objectification of private motor travel, a computer can be looked at as the objectification of the exercise of a certain form of control over a production process, and so on.

Beyond those activities which the individual can carry out without the aid of machines or tools – a sphere that has continuously diminished as technology has increased in importance – all other social actions require the manipulation of some form of machine or tool. We can therefore say that, at this level, the machines to which the individual has access coincide with the forms of possible social action that lie open to him. These actions include those which might appear relatively personal, such as cooking; not only does such an action rely on the activities of others – those who produce the food or the power used – but also the way that the activity is carried out, and hence the specific machines that are used in the process, tend to be held in common by all members of a society. Virtually our whole daily routine is achieved through machines. Shaving, making coffee, frying an egg, catching a bus, or a train to work, speaking on a telephone, watching television, each of these actions are achieved through the use of an element of technology. And the succession of machines that we employ can, conversely, be intepreted as the 'objectifications' of our sequence of daily activities.

Technology, in fact, can be considered as a 'language' of social action. From this language we select those individual elements – whether tools or machines – which we then apply to carry out a particular task. The instrumental (means to an end) activity of the individual, carried out through the successive applications of a variety of machines, might be considered as a personalized form of 'speech', corresponding to the notion of technique as it has been defined above. And in the way that language is experienced through the spoken or written word, which therefore become the 'objectifications' of social ideas and concepts, so, as we have seen above, technology represents in the individual machines of which it is composed the 'objectifications' of social action. Susanne Langer has written, in her book *Philosophy in a New Key*, that 'words are certainly our most important instruments of expressions, our most characteristic universal and enviable tools in the conduct of life'.[1] We can turn this analogy round, and suggest that machines increasingly provide the means by which social actions are organized and objectified as a language of action, from which elements are selected and utilized as required by an individual or a group in order to perform a particular task or activity.

While machines, like words, coincide with possible social actions that the individual is able to carry out, they also implicitly contain constraints on this action. Although we can carry out those activities in a social domain for which suitable machines are available, at the same time we are unable to carry out those activities for which machines are necessary but are not accessible. Some activities, indeed, become excluded by the introduction of a new machine. We cannot, for example, travel by train to those destinations from which services have been withdrawn, or, as I have discovered, be notified by a library with a computerized loan system immediately a book has become overdue, as although technically feasible, such possibilities are not incorporated into the design of the respective system or machine. Few of us are in the position to devise machines to perform tasks which we have selected or defined ourselves. For the most part, we have to rely on those machines which society makes available to us. The institutionalization of technology has meant that the choice of particular machines, or at least the control over this choice, remains in the hands of a dominant social class. And since technological innovation, as has already been suggested, is only carried out to the extent that it coincides with and maintains the interests of this class, new machines will only be introduced within the constraints that are imposed on the activities of the individual members of society.

A particular case of the possibilities and constraints expressed by technology are those contained in the forms of communication that can take place between different members of society. The design of media of communication determines the way in which it is possible for the individual to communicate with other members of society at anything more than a face-to-face level. The telephone system, for example, makes long-distance conversation possible between two individuals. As our present telephone system is organized, however, it excludes the possibility of group discussion without elaborate prior arrangements being made with the Post Office. The technical aspects of implementing this are, I have been assured, relatively simple. Yet the system was designed primarily for two-way conversation – the dominant mode of communication with an individualistic society – rather than group discussions. Another example of the constraints on communication implied by the telephone system are the financial ones. Those with sufficient means are now able to converse on a global basis. Those without these means are either denied the opportunity to use a telephone at all (at least not in their own homes), or if they attempt at all to use the system without paying, become treated as criminals. (The absurdity of the situation is shown up by the fact that those convicted at the 'phone phreak' trial in London in 1973 were made to face the charge of stealing electricity, a charge on which half of the accused were found not guilty.)

Other modes of communication, such as television or the railways, offer similar examples of how the articulation of a particular aspect of technology as a social institution already coincides with the way that it is intended the technology should be used. The railway system, for example, with its heavy emphasis on conveying businessmen between centres of production or commerce, and its systematic pruning of 'non-economic lines', immediately shows us what are considered to be the 'important' routes on which fast and efficient transport is 'necessary' (a point underlined by attempts by the EEC Commission in Brussels to get the railway authorities of different Common Market countries to co-ordinate their development plans aimed primarily at speeding up rail travel between its commercial centres). The railway system can thus be seen as an 'objectification' of one of the modes of communication required by a capitalist economy, one that requires the rapid transport of its business executives – and occasionally other classes, as in the case of migrant workers or those on holiday – yet is frequently able to ignore the interests of those less essential to the functioning of the economy, and indeed often to withdraw services to those in the name of 'rationalization'.

Another area in which the freedom allowed the individual, and the constraints imposed on him, become directly expressed in technology is that of industrial production. Within capitalist society, production is organized on the basis that the activities of the workers are subordinated to the decisions of management, these decisions being an expression of the interests of capital. The relationship between worker and manager is therefore a hierarchical and authoritarian one, with production ideally organized in a way that enables decisions to be taken at the top and carried out unquestioned at the bottom. Industrial technology not only provides, in its individual machines, the physical means by which decisions about the nature of his work are imposed on the individual worker, but also as formalized system – such as a production line – the means by which a general pattern of control is maintained over his work activity.

Within capitalist society the wealth produced by the productive activity of the worker is turned into capital, much of which is subsequently reinvested in technological equipment. This equipment is frequently used simultaneously to harness the worker to a particular type of work, to dictate the rate at which he works, and to achieve the reproduction and increase of the capital that employs him. Industrial technology therefore embodies the social relations between men required by a capitalist economy. 'The development of the means of labour into machinery is not fortuitous for capital, it is the historical transformation of the traditional means of labour into means adequate for capitalism,' writes Marx in his *Grundrisse*.[2]

We can extend the analysis of the technology employed in industrial production to the whole of the social world. Capitalist society relies for its reproduction on the rapid turn-over of commodities, and hence on the exercise by the producer of close control over patterns of consumption. In many cases, the machines purchased by the individual have been designed with these considerations in mind. We have, for example, the concept of 'built-in obsolescence' incorporated in a machine to make sure that they do not last for too long, or stylistic innovations that are used to identify 'next year's model' that will once and for all, we are told, put Mr Jones in his place. Thus even when looking at the machines possessed and used by the individual, we see that each machine serves to meet his apparent needs only to the extent that his 'consumption' or 'appropriation' habits conform with the behaviour that is required of him to maintain the effective operation of a capitalist society. Even the do-it-yourself movement, whether in house-building, car repairs, or whatever, has been turned by commercial exploitation into an extension of those industries.

In general, we can say that a society's technology, when viewed as a social institution rather than a heterogeneous collection of machines and tools, is structured in a way that coincides with its dominant modes of action and interaction. Although these modes of action and interaction are maintained and, often, reinforced, by technological factors, they are primarily a historical product of the power relationships within a given society, and hence of the economic factors by which these relationships are expressed. Yet the modes of action and interaction embedded in technology themselves coincide, as we have seen for example in the case of the telephone system, with the pattern in which power is distributed in society. Technology does not just provide, in its individual machines, the physical means by which a society supports and promotes its power structure; it also reflects, as a social institution, this social structure in its design. A society's technology can never be isolated from its power structure, and technology can thus never be considered politically neutral.

So far, we have looked at technology from a synchronic perspective that provides us with a sort of 'snap-shot' view at any particular point in time. Society, however, is in a continuous state of transformation and if we are to understand the role that technology plays in this process, we must refer also to our previous discussion of the process of technological innovation as the way that particular machines are incorporated into the functioning of society.[3]

I have previously argued the essential nature of technological innovation as a political process. In an attempt to guarantee the maintenance of its hegemony as a prerequisite to its material prosperity, and sometimes its very survival, the dominant class in any society – whether feudal, aristocratic or capitalist – will attempt to ensure that wherever possible technological innovation remains subordinate to these requirements. Although the increasing efficiency of the production process may be required by a capitalist society in order to ensure its *economic* survival, and this in itself be one determining factor in technological innovation, at the same time innovation has become an important means of attempting to ensure its political survival too. In the terms of our previous analysis, one of the purposes of innovation is to maintain and, where possible, to reinforce the dominant modes of action and interaction in any society, and hence the existing distribution of power. As long as technological innovation remains in the hands of a dominant social class, it becomes an important means of defending the power structure. And it is only, I would maintain, when the hold on innovation is wrested from this class as it was from the feudal landlords by an emerging merchant class, in the early stages of capitalism, that technology's potential as providing the means of sustaining a new form of power structure, or a new system of social relations of production, is released. In both the development of the factory system, and the subsequent technological innovation, for example, we see how the new patterns of social control of industrial capitalism became incorporated in the modes of production.

Beyond its merely economic function there are a number of ways in which the political dimension of technological innovation becomes apparent within the industrial situation. In particular, there are four situations – at least – that can be identified. The first is where technological innovation is used directly to increase supervision and tighten control over the activities of the work force. The instance quoted previously of the introduction of factory-based construction techniques in the building industry is one example. The second situation is where innovation has been used as a means of introducing stability into a work situation by displacing militant factions of the work force; here the introduction of containerization into the ports provides a typical illustration. The third situation is one in which the purpose of innovation is to achieve an apparent improvement in the work situation, and hence a removal of possible source of conflict; the widescale introduction of working methods referred to as job enrichment, where work is carried out in independent teams rather than in a linear production line system, as a typical example of this. By placing relatively insignificant decisions in the hands of the worker, such as the rate at which he decides to work to meet a predetermined target, it is hoped that pressure will be taken off demands on significant issues, such as rates of pay, or the level of targets. Finally, technological innovation can be used as a threat to blackmail sections of the work-force into particular tasks; a number of employers, for example, have claimed that if demands by women machine operators for equal rates of pay with men were granted, then the employers would introduce machines to take over the women's work.

Thus in both direct and indirect ways we can see how technological innovation serves a political purpose by maintaining and reinforcing the position of a dominant social class, and how technological development can therefore be interpreted as a political process. Again this illustrates the central thesis that technology, as a social institution, plays a directly political role in society. It demonstrates the manner in which technology coincides with the dominant modes of action and interaction, and can be seen as one of the essential mechanisms by which a particular set of social class relations is maintained.

So far, I have concentrated on a description of the objective functioning of technology in society. This analysis, however, would be sufficient only if all man's actions were purely instrumental in nature, in other words perceived and carried out merely as a means to an end. It would also require that the consciousness of the individual was developed through a relatively simple feedback mechanism from an awareness of the social world as a mere object, and from the effects that his instrumental actions have on this world.

We must now ask, however, how the individual perceives technology and his relation to it. We must deal, in other words, with the realm that covers the *significance* attached to technology as well as its objective role. I hope to indicate how this significance serves to maintain the cohesion of the social world in which we operate, and the social reality that we inhabit. It does this, I suggest, by providing us with an interpretation of technology that disguises the exploitative and alienating role technology plays within industrialized capitalist societies, and leads us to accept a particular mode of technological development as being a unique, inevitable and politically-neutral process.

This interpretation, I maintain, is experienced through one of the major myths of contemporary society, the myth that technology is, in fact, politically neutral. One of the functions of any myth is to explain or interpret a certain situation or set of events by providing them with an apparent social or historical context. this is used either consciously or unconsciously as a substitute for their real history, and hence removes such a situation or event from an awareness of its material genesis. Many myths, for example, such as the stories of the Old Testament, have formed an important element by which the social function of religion is articulated. Myths, in other words, are one important way in which any ideology is experienced by an individual.[4]

The myth of the political neutrality of technology is frequently invoked both by the politician, seeking to explain the effects that a particular technological development has on society, and by the technologist, seeking to disclaim responsibility for such effects. Problems such as industrial pollution, traffic congestion, or indiscriminate nature of military technologies, will be blamed either on inadequate policies or insufficiently sophisticated technologies. In each case the technology itself remains blameless. Scientists and technologists who have developed techniques that are subsequently used in military applications, such as defoliants in Vietnam or infra-red rifle-sights for night use in Northern Ireland, claim that the responsibility for the development of such techniques should be clearly identified from responsibility for their use.

The substance of the 'technology is politically neutral' myth is that the outcome of any particular act involving the application of an element of technology is determined entirely by the motives behind that act. The technology utilized is held to be entirely independent of such motives. It claims that we first select a particular task that we wish to carry out, and then select the appropriate machine which will enable us to perform this task.

Although the choice of the task may, according to this notion, be consciously or unconsciously determined by political factors, the subsequent choice of machine or technique to achieve the task is claimed to remain a politically neutral, technical task.

There are, of course, other myths that have been associated with technology in the past. Industrial power, for example, was an important symbol of Britain's imperialist strength during the nineteenth century, while heavy industry took on an almost mystical significance for the Bolsheviks in the early years of the Russian Revolution. There is also the contemporary myth that our social problems are in many ways *caused* by advanced technology, in other words, that advanced technology has become intrinsically anti-human; although this myth ignores the political factors behind the oppressive nature of technology in advanced capitalist societies, it nevertheless comes close in a number of respects to the interpretation I have suggested in the previous section by refusing to separate the nature of technology from the uses to which it is put.

The myth of the neutrality of technology however legitimates the contemporary social function of technology by attempting to take it out of the realm of political debate. We cease to question the nature of the institutionalized structure of technology and the way in which it has developed historically, and accept both as natural, or rather as inevitable. Thus we are told that the railways are run in a certain way because it is the most 'efficient' way to do so, or that the *only* way of getting over current environmental problems is by increased technological sophistication.

Furthermore, this myth is an essential element of a general mechanistic and functionalist ideology of contemporary society. This ideology postulates an equivalence relationship between the concepts of industrialization, modernization, and technological and social development, already referred to as the 'ideology of industrialization'. It is this ideology, for example, that legitimates the continued pursuit of economic growth, placing primary emphasis on the generation, rather than the distribution, of wealth. It also seeks to explain the functioning of society purely in operational terms, denying the relevance of political considerations concerned with the distribution of power or patterns of social control. Within industrialized capitalist countries – and frequently in the underdeveloped countries too, when a nascent capitalist economy has been enticed into a world market – it is used to suppress the potentialities for individual and social emancipation offered by particular machines and legitimate their use towards socially-exploitative ends.

The same myth provides an interpretation of technological innovation as the objective, technical response to particular circumstances. Here the legitimation can be expressed in terms of 'increased efficiency' (e.g. the introduction of assembly-line production techniques), as the 'technical' solutions to social needs (e.g. the development of a new transport system) as the economic 'rationalization' of out-date techniques (e.g. the introduction of automatic telephone systems) or even as the need to maintain a competitive position in a world market (e.g. Concorde). Whatever the terms used, they attempt to imply an objective rationality to the process of technological innovation that presents any particular innovation as being the 'logical' response to a given objectively-defined situation.

Already it can be seen that many of the terms and concepts utilized by such an interpretation have been derived from the practice of science. Indeed, it is possible to suggest that the legitimation of both the social function of technology, and of technological innovation, is based largely on the ideology of 'scientism', which is itself legitimated by the positivist interpretation of the practice of science as the pursuit of an empirically-discovered objective truth.

Positivism as a natural philosophy combines the traditions of empiricist and rationalist thought. This was originally done to strengthen the belief of scientists (and frequently of non-scientists too) in the exclusive validity of scientific truth. Scientism is the transformation of positivism into a social philosophy, the basis on which man explains and interprets the nature of society. As such, I maintain, it forms an essential element of what has been referred to as the ideology of industrialization. In essence it conveys the conviction, as Jürgen Habermas expresses it, that 'we can no longer understand science as one form of knowledge, but must rather identify knowledge with science' (in his *Knowledge and Human Interests*).[5] The basic message of scientism is that an apparently 'scientific' approach to any problem or situation is both necessary and sufficient to indicate how its objective, politically-neutral resolution can be achieved.

Furthermore, it can be argued that the importance of scientism as part of a contemporary ideology is closely associated with the heavy dependence of advanced industrialized societies on discoveries made in the scientific laboratory. In traditional tribal societies, technology was derived directly from human practice, and this practice was implicitly related to the dominant values and social patterns of the prevailing culture. The very scale and complexity of modern technology, however, precludes the possibility of its development from direct collective experience. Technological development now relies heavily on the application of abstract modes of reasoning, and on knowledge frequently – although arguably – attributed to the results of scientific investigation.[6]

Technological innovation does not spring directly from a laboratory bench. The application of abstract knowledge to practical tasks, as we have seen above, is already a politically-determined process. The legitimation of this is provided by the apparently neutral interpretative scheme of positivist science, rather than collective practical experience. It is the legitimation that science provides, as much as the potentially-useful knowledge that it produces, that is important for an understanding of the social role of contemporary technology.

This process can be seen at work in the development of ideas of scientific management. The division of labour that emerges, under the application of scientific management, from the supposed scientific analysis of the practice of commodity production becomes embedded in the design of production technology. The way a factory production line is designed, for example, implicitly contains the idea that production line methods are the most appropriate for a particular mode of production. The social fragmentation of the work-force necessitated for the efficient running of a hierarchical system becomes directly experienced in the physical space placed between each worker on the shop floor. Such innovations are justified, however, on the grounds of 'increased economic efficiency' or some such argument, implying that it has been possible to mathematicize the work situation in an objective, scientific way. Scientism thus legitimates a process by which the division of labour, initially emerging as an economic and social necessity, has subsequently become institutionalized in the very design of production technology, while at the same time being fashioned to meet political ends.[7]

The same situation occurs outside industrial production. Edmund Leach, for example, has suggested how science is used as a legitimating factor in contemporary medicine. 'Medical knowledge has ordinarily been acquired and preserved as a technology; a practical way of dealing with a particular situation,' he writes. 'It was only quite late in the present century that it came to be generally accepted by the medical profession that innovations in medical practice ought to be justified by scientific criteria.'[8]

In social planning the use of supposedly-objective models of human and social behaviour serve to legitimate the imposition of social policy. Development planning in the underdeveloped countries has similarly been predicated on a model of Western positivistic rationality which seeks to imply that the social problems of these countries can be solved by the rational application of Western science and technology, and by fostering the development of Western-style institutions, attitudes and values.

The use of the 'scientific method' in this way makes it important to refer briefly to the origins and development of modern science. It was during the philosophical revolution which took place in Western Europe in the sixteenth and seventeenth centuries that the general validity of a secular mechanistic science as a natural philosophy began to be accepted, not just by a few isolated intellectuals, but on a broader social scale. Breaking away from the essentially non-functional traditions of Greek science, the first modern scientists, including figures such as Copernicus, Galileo and Francis Bacon, began to realize the potential value of scientific knowledge to society as providing it with the means, not only to interpret the world, but also to change it.

The philosophical basis of this new science was provided by Descartes, whose strict division of reality – or rather of knowledge of reality – into the categories of subject and object have provided the basis of the practice of Western science for the past three hundred years.

Descartes' method was not the only one available. The philosopher Leibniz postulated an alternative interpretation of reality in his *Monadology*.[9] His philosophy of monads was based on concepts of wholeness, and emphasized the relationships between phenomena and between events, rather than the distinctions which separated them. Yet this precluded the possibility of a scientific method that would allow men direct access to the forces of nature, and was hence unable to ally itself to an instrumental material base. It was Cartesian logic that held the day; Leibniz's ideas were allowed to fall into obscurity. One consequence of this was the apparent social acceptance of a cleavage between values – the realm of the subject – and facts – the realm of the object. This in turn appeared to legitimate the separation of feeling from thought, a process that can be traced in almost all fields of cultural activity. After the time of the metaphysical poets, to take but one example, we can recognize a sharpening of the distinction between poetry and prose. Rousseau later claimed that 'Descartes' philosophy had slit the throat of poetry' (*'la philosophie de Descartes avait coupé la gorge à la poésie'*).

This whole process was accompanied by the emergence of the notion of purity applied to various forms of cultural activity. Science despite recognition of its instrumental importance became 'pure' science, consciously divorced from its social applications and frequently justified in its own terms as the disinterested pursuit of knowledge. The idea of art for its own sake (*l'art pour l'art*) appeared at roughly the same time, first in the busy commercial climate of fifteenth-century Italy, and subsequently in seventeenth-century France under the politically-motivated encouragement of Louis XIV. During this and subsequent periods, the myth of the intrinsic value of the art object was used by an emerging merchant and capitalist class to legitimate the appropriation of wealth by objectivizing it in non-commercial, i.e. cultural, commodities.[10]

We even find the concept of pure work beginning to emerge at a roughly similar period with the beginning of the wage system of labour. Labour-power was now something that had to be exchanged, and this necessitated the dissociation of the worker from his social context, in other words his alienation. Originally subject to collective group relationships, themselves responsive to the natural rhythms of social intercourse, work became regarded as an economic commodity as it was gradually drawn into the organization of factories, where it was directly subject to principles of efficient organization of production. This process was assisted by those religious movements such as Methodism which preached that hard work was, *per se*, a virtuous and soul-saving activity.

The idealistic concept of the intrinsic value of an individual's activity, whether mental or manual, divorced from its social role hides an important paradox. Advanced industrialized societies tend to be characterized by a strong moral and cultural emphasis on the value of abstract scientific knowledge – even that with no immediate or apparent application in sight – and on the apparent need to relentlessly push back 'the frontiers of scientific discovery'. Yet the fruits of the application of this knowledge to practical tasks are an essential element in the efficient operation and expansion of such societies. Whereas in a cultural domain, the abstract, almost metaphysical qualities of science squeeze out the more mundane instrumentality of technology, within the sphere of social production the relative importance of the two is dramatically reversed. In other words, the stress placed on the cultural importance of abstract science legitimates the ideology of scientism, yet disguises not only the exploitative way in which science is put to practical use through technology, but also the very fact the existence of contemporary science – in terms of support for R and D – results directly from this practical use.

A further aspect of scientism is that it promotes a passive acceptance of an existing state of affairs. The function of scientists becomes to provide an explanation for why it is 'natural' that things should be – and thus remain – as they are. Scientism, by accepting such explanations, stimulates a general amenability to manipulation by political forces. Science may contain the potential of being transformed into a materially progressive force; but under the ideological disguise of scientism, it seeks to obscure the possibility of political emancipation, and is used to suppress such a possibility by preaching the 'natural order' of the existing system. It dismisses as irrational or unscientific any attempts to challenge our contemporary situation in terms of the class interests which it maintains. (A group of correspondents to the *Guardian* in April, 1973, drew attention to a news story reporting psychological 'research' as indicating 'sexual attractiveness as a predictor of attitude and response to risqué humour' by looking at reactions to 'dirty' postcards. They write: 'This is surely the most blatant use of the label of "science" to legitimize what is actually prejudice, stereotyping and denigration of women'.)

The officially-sanctioned view of the world becomes one seen through the 'objective' categories of science. Thus we see, for example, the Roskill Commission on the siting of London's third airport taking into account only those social factors which can be 'scientifically' quantified; or behaviourist psychologists reducing the cultural and social differences between ethnological groups to measures of performance in an 'intelligence' test. Economists tell us that if something exists, then we can quantify it; sociologists and psychologists conversely tell us that if we can quantify a concept (such as intelligence) and show that it behaves in a quasi-scientific way, then this is sufficient to indicate that the concept exists in an objective fashion.

It is not difficult to find expressions of this ideology. A typical illustration is provided by Victor Ferkiss in his book *Technological Man*, when he writes that 'technological man, by definition, will be possessed of the world view of science and technology, *which will themselves provide a standard of value for future civilization*'.[11] Others such as the French biochemist Jacques Monod have claimed to demonstrate that science itself provides a rationality that has superseded, and hence can be considered superior to, previous forms of religious or political rationality.[12] Scientific truth becomes the substitute for religious truth. Yet its ideological and dogmatic nature – by definition imperceptible to those who believe it – is no less dangerous than that of religion. For by attempting to set up a system of rationality that is apparently *above* man, it distracts attention from the way that such a rationality can be used to obscure the nature of the relationships *between* men.

Furthermore, by maintaining the Cartesian split between the abstract world of the subjective and the material world of the objective, the ideology of scientism legitimates the idea that conception of a particular line of action can and should be divorced from execution, and theory from practice. Whereas Marx suggested that human labour was a value-creating process that integrated the essential elements of social life, and in particular brought together the realm of the subjective and that of the objective, scientism seeks to legitimate the division between for

example, 'cultural' and instrumental activity, and thus to maintain the confrontation between subject and object.

We see the false distinction between objective and subjective
reality reflected in what is referred to as the 'two cultures' debate,
initiated by C. P. Snow in 1959. Although it may not have been
Snow's original intention, this debate has now come to signify the
confrontation between the sciences on the one hand, and traditional human interests on the other, that appears to manifest itself
in many aspects of social life. What the debate generally fails to do,
however, is to distinguish the essential nature of an activity from
the way in which it has been institutionalized, and subsequently
mythologized, as part of the dominant ideology. The reasons for
the apparent contradictions between the sciences and the humanities are to be found in an ideology that first identifies these as
autonomous activities independent of the material base of society,
and having done so, gives them opposing roles in a way that
virtually guarantees the emergence of conflict.[13]

In the above two sections, I have indicated, firstly how technology
functions in a political fashion in society, and secondly how its
political role is disguised by the dominant ideology of the system.
This is done through the existence of the myth of the neutrality of
technology, and the interpretation this provides of both the social
function of technology itself, and the process of technological
development. In this section, I hope to indicate some of the consequences that this interpretation of the role of technology, and
its basis in the legitimating ideology of scientism, has for the way
that we interpret the nature of technology. I want to suggest that
in a symbolic sense, technology appears as an inevitable or 'given'
element of the social reality that we inhabit. This reality appears
divided into separate ways of looking at events or phenomena,
for example from a political, a historical, a cultural or an economic
perspective. Technology, or rather the objects that it comprises,
appears to play a neutral role within each perspective, as the mere
expression of a reality that lies beneath it.

It is this situation, I suggest, that attempts to persuade us to
ignore the ideological nature of our interpretation of the function
of technology, and accept technology as a 'natural' element in our
social reality. In other words, not only does technology function

politically in a material sense to promote the interests of a domin
ant social class by maintaining a particular form of social organiza
tion; it also, I suggest, functions politically in a symbolic sense, in
that the particular interpretation placed on technology is one
designed to support a dominant ideology. Because the individual
is conditioned to accept a socially-given interpretation of the world
and the place of technology in it as being natural, and hence in
evitable (through, for example, the reinforcement provided by
the educational system or the mass media), the political nature of
this interpretation is obscured. Furthermore it is obscured within
each apparently autonomous perspective on reality, while it leads
us to interpret the instrumental activity of the individual as the
consciously-designed acts of a socially-autonomous, rationally-
motivated subject, rather than itself a process which is subject to
the societal constraints imposed on the individual on the one
hand, and the expression of his own physiological motivations on
the other.

The importance of looking at the way technology functions
within different perspectives, or what I shall call 'modes of significance', springs from the fact that an individual machine performs many roles in society. A car, for example, can function for
its owner not only as a means of private transport – its direct
material function – but also as a social status symbol, a week-end
hobby, a means of self-expression, a desirable object, a business
reward (in place of a salary increase) or even, if suitably antique,
as an economic investment. A recent advertisement, for example,
implied that the Italian gliding down the Autostrada del Sole at
200 kilometres an hour in his Rolls-Royce 'is fulfilling some important aspects of his nature. His love of beautiful things. His
appreciation of engineering technique. His passion for driving.
And his pride'.

For the car manufacturer, the car functions simultaneously as a
profitable commodity, as an expression of his apparent contribution to the development of society, or as a determining factor in the
design of his production-line or his sales strategy. For the non-
owner, the car can become a pedestrian hazard, an environmental
menace, a direct challenge to public transport, and, as a spectacle,
a sublimation of unconscious drives. To those living in underdeveloped countries, it can function as a status symbol, a desired

by-product of modernization, a source of employment – either in factories or garages – a drain on foreign reserves, or a symbol of economic exploitation by neo-colonialist forces.

The same machine functions simultaneously in many different ways – or one might say expresses a number of different values. Many of its functions differ from those which the machine was initially designed, in an instrumental sense, to fill. The model T Ford, for example, was in no sense 'designed' for its antique value sixty years later. The total symbolic function of each machine is therefore a composite one. Although it may originally have been developed to perform a specific instrumental function, as the means to a particular end, its existence within a social environment – whose importance to the individual goes far beyond the material properties of the objects of which it is made up, but provides him with the framework for his whole interpretation of reality – endow it with a range of further functions.

One can no longer say that *the* function of a machine is to perform a single, isolated task. Each performs a range of functions. And in many cases these functions are purely a product of a social environment – cars as antiques, for example – and totally unrelated to any material role. Conversely to condemn a machine merely on the basis of one of its roles – cars as a source of pollutants, for example, or computers as a source of clerical error, is shallow, even if deserved, criticism. It is misleading to separate one role of a machine from its others in this way, just as we have previously seen it to be misleading to separate the action of one machine from that of another in assessing the function of technology as a social institution.

The essential task is to discover how one particular role both relates to and reflects the 'values' of the social milieu in which it has been defined, and of which the machine has come to form both an extension and an integral part. To do this, it is necessary to re-establish the connections which have been severed between what I have called the different modes of significance, in order to see the extent to which the articulation of 'meaning' within each is itself a socially-determined process. We tend, for example, to separate the apparent economic value of an object from its apparent aesthetic value, without realizing the extent to which both are expressions of a particular way of looking at the world, itself the product of our social experience within a class-based society.

Taking the above illustration of the motor car, we might say that when interpreted as an economic investment, a business reward, a profitable commodity, a challenge to public transport, or a drain on foreign reserves, we are referring to its economic function or significance. As a status symbol or a symbol of foreign oppression, it is functioning in an immediate social or political fashion. Interpreted as a source of pollutants or as an object requiring an amount of non-renewable natural resources for its construction, we are looking at its environmental significance. Yet it is not difficult to see how the meanings associated with it at each of these levels – such as a drain on foreign reserves becoming a symbol of foreign oppression – are closely associated to one another. Reality, in other words, forms a 'seamless web' within which any fragmentation or categorization is, it can be argued, a socially-determined process.

The general acceptance, however, of a clear differentiation between each mode of significance means that each mode is treated as being defined independently of any social determinants, and subject merely to the articulation of an internal logic (such as, for example, the so-called laws of classical economics or of aesthetics). Technology, as we have indicated, functions symbolically within each of these modes of significance, in other words provides a set of symbols by which the apparent significance or value of a social phenomenon can be articulated. As in the discussion of its material significance, however, technology tends to be treated as if it were a neutral element in this process, a symbolic tool by which significance is expressed, rather than essential element in the process of expression itself.

Take, for example, the relationship between technology and another conventionally-accepted mode of significance, that of art. We have already referred to the ideological nature of the 'two-cultures' debate. A particular aspect of this debate is that, rather than talking about technology and art lying within the same dimension of collective experience, technology is referred to as something isolated from, and even, to a certain extent, opposed to, man's 'artistic' activities. This is apparent in debates that take as their terms of reference the 'impact of technology on art' or the apparent confrontation between man the artist and man the technologist. We read discussion on technology and art, or technology in art, but seldom technology as art. It is considered legitimate for the artist to use technology in his work, but only as a means of expressing his (artistic) ideas. Kinetic art and computer art provide two examples of this. In such instances, the artistic and technical aspects of any work of art produced in this way usually remain clearly defined. Not only does technology appear politically neutral, but it appears neutral within the artistic process too.

A second, possibly more important, level of significance is that of political debate. We have already seen how the myth of the neutrality of technology obscures the political function of technology in society. A corollary of this is that attempts are made to neutralize any political opposition to the use of technology by translating it into the de-politicized language of the myth itself. This is the case with the so-called 'technical fix'. Here it is claimed that the social problems apparently created by technology can be solved by the application of yet more technology. We are informed, for example, that the use of sophisticated monitoring and control devices will itself be sufficient, given adequate government legislation to ensure their use, to combat the problems of environmental pollution, or of appropriately-designed technologies to solve the social and economic problems of the underdevelopment. The argument is not against the necessity, or even the adequacy, of the individual machines implied by such policies, but against a scientistic analysis that both ignores the political roots of the problems that it seeks to solve, and subsequently pretends that solutions can be achieved merely through technological means.

The same is true, although in a slightly more subtle way, of the development of techniques of technology assessment. The question here is not whether the impact of technology on society should be assessed, but how it should be assessed. As currently being developed within industrial capitalist societies, technology assessment seeks to evaluate this impact – and hence to recommend those measures by which the 'undesirable' aspects of technological innovation can be minimized – using the very same methodological categories (such as objectivity and political neutrality) as those by which the technology itself is claimed to have been developed. It obscures the fact that the nature of the categories selected as the basis for analysis already disguises the essential political elements of the problems under review.[14]

The converse of the neutralization of criticism is that those who propose technological programmes in the belief that such programmes are outside the realm of political debate, while prepared to listen to criticism of a purely technical nature – and frequently to modify plans in the light of such criticism – are nevertheless able to justify the dismissal of any *political* challenge as being irrelevant to the issue under discussion. Thus an official inquiry into the siting of a motorway will concentrate on where such a motorway should be placed to create the 'least' environmental or social disturbance, not whether a motorway is really needed at all, or whose interests the motorway will serve.

Related to the de-politicization of technology, of course, is the interpretation of politics itself as a purely technical activity, and the evaluation of political activity in primarily instrumental terms. Political programmes tend to be assessed on what they have achieved in terms, for example, of economic performance. Political parties are compared on the effectiveness by which they have been able to handle a capitalist economy. The extent of aid to under-developed countries is evaluated in terms of its size as a percentage of gross national product. Thus the convergence theory of politics, the alleged death of ideology – since everything is now claimed to be done 'scientifically' – and the apparent development of a politically-neutral technocratic society.

This process also takes place in education, where techniques such as the Keller-plan form of programmed learning, based heavily on Skinnerian behaviourist ideas, are again judged purely by their effectiveness at improving the speed and efficiency of what becomes known as the learning process. The fact, frequently quoted by supporters of such techniques, that students 'prefer' such methods of teaching since it is more effective in helping them -to pass exams is itself an indictment of the nature of the educational system and the forms of assessment that it uses, rather than any indication of the intrinsic value of the techniques. It is these and similar facts that lead to the characterization of contemporary ideological legitimations we have referred to previously as 'mechanistic' and 'functionalist'.

To summarize, I have tried to indicate how technology functions in society not merely in a material sense, but also in a symbolic one. It does so within a number of what I have called modes of significance. The conventional interpretation of this process, I maintain, is that the symbolic function of technology within each of these modes is a purely neutral one: the symbol is considered independent of what it is supposed to express. Furthermore, each mode of significance, whether economic, cultural, political or even – I would claim – subconscious, is held to be independent of the others, and subject only to its own internally-articulated logic. This whole interpretation of the social world in which we live, I suggest, disguises the implicit connections between the various modes of significance. It also hides the role that technology (symbolically) plays in disguising, while at the same time reinforcing, the apparent nature of each of these modes of significance and the extent to which they are politically-determined, through maintaining the myth of its neutrality. Technology functions symbolically in an ideological fashion to maintain and reinforce a particular form of social organization and control just as it functions materially towards the same ends.

The purpose of this discussion of the social function of technology has not been so much to provide an explanation for a particular state of affairs as to describe the legitimation that a dominant social class attempts – not always successfully, one should add – to place on that set of affairs. In no way, therefore, does it lead directly to a recipe for an adequate political practice that would alter such a situation. Yet it does provide some indications of various aspects required by such a practice. In particular, it points to the need to challenge both the uses of technology and the process of technological innovation directly on political grounds, as part of an overall struggle for emancipation from the exploitative system which technology maintains in industrial capitalism.

One suggestion has been for the development of a 'science for the people', by which the scientist or technologist places his skills and knowledge at the service of the oppressed groups of society in order to redress the monopolistic control over the practice of science and technology at present held by society's dominant groups and institutions. A similar activity, suggested by the British Society for Social Responsibility in Science, is 'community science'. The two primary aims of community science have been suggested as being to provide scientific and technological services to the community – such as advice on technical aspects of health care – that are largely neglected or unrecognized by the dominant ideology, and the institutions on which it is based, and to develop those aspects of the class struggle that involve some scientific or technological component.[15] Examples of the latter might range from challenging the so-called experts who are employed by the major institutions of both industry and Government to provide legitimation for oppressive policies, to direct development of the technical means for assisting in class struggles. As two members of the US Black Panther movement, both research scientists, have explained 'It's a question of what the community needs. If a person knows that he can make a contribution through maths and can see that there is a need, then Right On . . . As many of our people as possible should learn technology with the idea of countering the technology that the Government is exploiting against us.'[16]

Such activities are important to the degree that they reintegrate theory with its practical applications, help to demystify the concept of expertise of the scientist or technologist and eliminate the deference that accompanies it, and act generally to raise the level of consciousness about the nature of society for those directly involved. The category of community science, however, itself implies a certain contradiction, for by removing science from the laboratory, one is already breaking the codes that define the activity of science, and hence throwing into question the very label 'scientist'. To retain this label is to retain the structure of expertise and mystification that characterizes industrialized society and the whole mythology of the neutrality of both science and technology; to reject it is to reject the identification of a category of activities that can be labelled as community science and can be distinguished from other forms of community activity.

The label community science implies, to a certain extent, that it is possible to identify science as an activity autonomous of its social and political context, and that the problems are merely related to the nature of the set of linkages between science and society. Such an approach denies the extent to which ideological factors are constitutive of the internal workings of science. Furthermore, it creates the danger that it might easily share the fate of, for example, community medicine and large parts of the radical housing and ecology movements, which have been co-opted into the dominant ideology and rapidly lose their critical edge. The important thing is not so much a 'science *for* the people' as a 'science *by* the people' – or the 'socialization of the intellectual means of production' – with all the implications for the distribution of power that this contains.

This is not to deny the importance of community-based activities. Any political struggle, whether it is a localized community action or a national movement, can help to counter the myths initially perpetuated by the educational system, and subsequently maintained by the operation of the mass media. Evidence for this can be found in the way in which the gross misuses of science and technology both in the development of nuclear weapons, and the development of other forms of technological warfare in particular by the United States in Vietnam, have awakened the idea of the 'social responsibilities' of the scientist, itself a precursor to the general process of raising a political consciousness. Any extension of the political debate outside the purely industrial dimension can lead to an awareness of the extent to which political factors pervade all aspects of daily life, both at work and in the community.

There are, too, purely practical functions that scientists and technologists can fulfil in the community. Alternative technology can provide a safety net for those whose needs the system fails to meet. The development of squatting – placing homeless families in empty houses which can be made habitable with a limited amount of technical knowledge – is perhaps a good example of this. Similarly a group of 'radical technologists' in Sheffield has been applying elementary skills in electronics to set up intercom systems for the housebound. The level of skill or knowledge required may not be very high, but this is less important than the positive function played in stimulating community cohesion and demonstrating the viability of 'self-help' strategies that do not rely on a centralized system of servicing or decision-making. Equally, there are opportunities for research scientists to help improve the working situation of those in industry. 'Why shouldn't working students in medicine, physiology and psychology begin to apply such experiments on a large scale to their own experiences in a modern enterprise, above all to description and analysis of the experiences of their fellow workers?' Ernest Mandel has asked. 'Critical medical students will be able to analyse the problems of fatigue, or frustration caused by alienated mechanical labour, by a steadily rising intensity of labour, better than positivist doctors – if they combine real professional expertise with a grasp of social phenomena in their own context, and enrich this with personal experience.'[17]

Such activities help to reveal the ideological nature of the myth of the neutrality of both science and technology. There is a danger contained in this approach, however, and that is that it can act as a personal rather than a social or political alternative for the scientist or technologist. By concentrating on localized issues and isolated problems, the approach can fail to locate the true centres of political power, or create an awareness of the structural causes of the problems involved. Frequently community activists have found that once a short-term objective is achieved – for example the provision of a play area or the removal of a local pollutant – the general impetus for change also stops. Similarly, it is often assumed that once viable alternatives, whether they be in terms of community structures or alternative technology, have been constructed and their viability demonstrated, then the existing system will merely dissolve or disappear (perhaps in the way a dying elephant appears to desert its herd or a snake shakes off last year's skin). It is claimed that the integrity of these alternatives, and thus the duplicity of that which they are destined to replace, will become self-evident, and society will attain a state of perfection through a continuous process of evolution. Once man has rediscovered his innate common-sense, we are told, he will be in a position to control technology rather than being controlled by it, and the path to a technological Utopia will be clearly and unambiguously defined.[18]

This analysis ignores the nature of the system of power that maintains our political structures. It also ignores the distinction that we have made between the present social practice of science and technology, and the mythology which disguises the nature of the practice. The power that resides in the major social institutions, and in the hands of those who control them, needs to be confronted, and the nature and function of the mythology that disguises this power must be revealed, before we can think in terms of setting up any alternative system.

In particular it is important to challenge the educational institutions, and the matrix of knowledge that they propagate. Both in formal and informal education, it is necessary to introduce an awareness of the role played by science and technology in modern society, and show how these relate to the nature of technology and the activities of engineers and technologists. It is important to attack curriculum structures that maintain patterns of social hierarchization, and support the mystique of the expert, legitimating the dominance of technical expertise.[19] The division of theory from practice in the teaching of science and technology must be challenged; this does not mean more of the conventional 'practicals', but a linking of science and technology as they are taught in the school or university with an awareness of the way in which they relate to the social experiences of the community outside, and of the practical ways in which they can help to develop these experiences. Finally the barriers that separate science and technology from the humanities must be revealed as products of the educational system and of the political system that supports it, and not as a natural division of the categories of culture or social activity.

The rapid expansion of science and technology since the Second World War has had two important consequences. The first has been the so-called proletarianization of the research worker, who finds his activities more and more controlled by the major institutions of society. A second consequence has been the increasing importance of hierarchies among research workers. The internal organization of research laboratories tends to reflect more and more that of hierarchical systems of industrial production, and challenges any traditional idea of the homogeneity of the research community.

The changing role of the research worker has had a profound effect on the significant modes of action that are open to him. Since he is neither in control of his own activities, nor of the way in which his research results are used, it is unrealistic to locate the concept of social responsibility at an individual, voluntaristic level. Control can only be regained through collective action, reflecting the nature of research as a collective activity. Marx pointed out in

Capital that manufacture 'is completed in large-scale industry, which detaches science from labour, making of science an independent force of production, and pressing it into the service of capital.'[20] It is only recently, however, that the full realization of this has come home to research workers themselves. A direct result of this growing awareness has been the increasing radicalization of research workers, and the establishing of collectives in research laboratories in the United Kingdom, France, Italy, the US and elsewhere, in an attempt by research workers and technical staff to regain control over the activities of the laboratory in which they work.

At the same time, however, the legitimating role of science and its implications for the design of technology cannot be tackled solely within the context of the research laboratory or the industrial firm. Any internal challenge to a mythology can be neutralized by translation into the language and categories of the myth itself. The social problems associated with the practice of science and technology cannot be solved merely by a re-assessment of research priorities within institutions as they now exist, but only by challenging the social role of the institutions themselves, and the way in which this determines both the nature of the activities carried out inside them and the way in which research priorities are decided, as well as the class status of those in whose interests these decisions are made.

The alternative to the current practice of science, writes Robert Young, is a 'critical and transcending view of science, one which looks hard at its reifications, its fetishisms, its role in alienation, and indeed the whole scientistic programme'.[21] The same can be said of the need for an alternative to the technology that is legitimated by this programme. As long as our present institutions exist, it is important to stimulate criticism from outside as well as inside, in order that the ideological role of these institutions may be revealed.

It is interesting to note here a reversal of the position held by pre-war groups of radical scientists. These were concerned to argue the case for a greater emphasis on the social importance of the scientific method, and to demand greater social status for the scientist, something that could be done from *inside* the scientific institutions. The opposite now appears to be happening, with scientists arguing *against* the extension of the scientific method as it is now practised into inappropriate spheres of social activity, and experiencing the contradictions that have led many to drop out of science altogether. Harold Wilson's *volte-face* between his 'technological revolution' speeches of 1963, and his 'man oppressed by technology' attitude of 1973, has been even quicker.

From the political point of view it is important to emphasize that the problems associated with advanced technology cannot be framed merely in terms of economic categories, concerning solely the ownership and control of the means of production, but challenge the political nature of our social and cultural institutions, that of the concept of the nature of man to which they have given rise, and the technological practices which have been based on them. Institutions that promote social hierarchies must be confronted with demands for the recognition of the equality and shared collective experience of *all* men. Not only must the division of society into oppressors and oppressed be broken down, but so too must the barriers that separate mental activity from manual labour, and abstract theory from concrete practice. Only through such changes can we create a situation that will enable us to reintegrate all aspects of social life and experience and to establish a situation in which man can be liberated to fulfil his full potential as a sensitive, creative and social being.

Of course, all this will inevitably have important implications for the development and design of technology. It can be argued, for example, that complex advanced technology is *per se* incompatible with systems of self-management, on account of both the social organization and technical expertise required to run it effectively. Similarly, direct ecological considerations and the limits to material growth seem destined to have an increasing effect on the choice of technology that will allow society to continue to enjoy the fruits of a stable yet productive environment. Certainly a technology that avoids the type of alienation discussed in previous chapters will need to be based on a different relationship between man and man, with direct consequences for the design of the machines themselves. The organization of production, too, must take account of the collective nature of work to assess optimal systems compatible with human – rather than economic – ends, and to serve the interests of all men, not merely those of a privileged élite. Yet in many ways as I have suggested it may be too soon to predict precisely the nature of these new machines and techniques. Many can only emerge from the process of political confrontation, and the resultant progress of social development.

Only by realizing the extent to which technology provides an integral part of the ideology of contemporary society, at the same time as forming an essential element of the mechanisms by which the supremacy of existing political systems are maintained, can we see the extent to which the need to develop an alternative technology is both necessary and desirable. To neglect the political dimension of this change is to support an idealistic concept of technology that does not coincide with the social reality of technology as it has been experienced. Yet to imply that the problem involves merely the social relations of production, and not the very nature of the means of production, is to disregard the extent to which our current technology is permeated by the exploitative ideology of advanced industrialized societies, whether nominally capitalist or socialist. Only when we have created a viable *political* alternative shall we be in a position to perceive the real needs for a community science or an alternative technology. But perhaps by then contemporary forms of science and technology as we know them will have disappeared, and the task of *consciously* setting out to develop such alternatives will become unnecessary.

References and Notes

Introduction

1 Jacob Schmookler, *Patents, Inventions and Economic Change; Data and Selected Essays* (Cambridge (Mass.), 1972), p. 81

2 Descartes, *Discourse on Method and Other Writings* (Harmondsworth, 1968), p. 78.

3 F. W. Paish and A. J. Culyer (eds.), *Benham's Economics* (London, 1973), p. 152.

1 The Case Against Contemporary Technology

1 There are, of course, numerous definitions of technology. The *Concise Oxford Dictionary*, for example, defines it as 'science of the industrial arts', while J. K. Galbraith, in his *The New Industrial State* (Harmondsworth, 1967), uses the word to refer to 'the systematic application of scientific or other organized knowledge to practical tasks'. I have chosen to use the word in a broader social sense in a way that suggests how the relationship between technology, machine and technique might be seen as roughly equivalent to that between language, word and speech (see Chapter 7).

2 Barry Commoner, *The Closing Circle: Confronting the Environmental Crisis* (London, 1971), p. 151.

3 Commoner, *ibid.*, p. 177.

4 Commoner, *ibid.*, p. 186.

5 John P. Holdren and Paul R. Ehrlich, 'One-Dimensional Ecology Revisited,' *Science and Public Affairs*, Vol. XXVII, No. 6 (June, 1972).

6 John Maynard Keynes, *Essays in Persuasion* (London, 1952), p. 372.

7 'Technological change has been the mainspring of economic and social progress over the past centuries and . . . it remains the chief source of our increasing affluence.' Sir Alec Cairncross, Presidential Address to the British Association for the Advancement of Science (September, 1971), p. 2.

8 'A company's capability in technological innovation can be defined as its ability to exploit science and technology in order to make or use profitably new or improved products and production processes.' Keith Pavitt, in 'Technological Innovation in European Industry: The Need for a World Perspective,' *Long-Range Planning* (December, 1969), p. 8.

9 See, for example: R. Clarke, *The Science of War and Peace* (London, 1971). Also S. Rose (ed.), *Chemical and Biological Warfare* (London, 1968).

10 Quoted in *Unemployment: An Oxfam Special Report* (Oxford, 1972), p. 3.

11 Judith Hart, *Aid and Liberation* (London, 1973), p. 42.

12 Mr Arthur Hawkins, chairman of Central Electricity Generating Board, told a Commons Select Committee that matters concerning the safety of nuclear power stations were 'best dealt with by experts on procedures laid down by governments and not debated before decisions are made.' (*The Guardian*, December 19, 1973).

13 Quoted in Noam Chomsky, *American Power and the New Mandarins* (Harmondsworth, 1969), p. 24.

14 Ralph Lapp, *The New Priesthood* (New York, 1965), p. 1.

15 Lapp, *ibid*, p. 2.

16 See, for example, Galbraith, *op. cit.*

17 See, for example, *Science Against the People: The Story of Jason* (published by Berkeley SESPA, 1973).

18 See: Joseph Hanlon, 'The technological power broker,' *New Scientist* (February 15, 1973). This seems more a reflection of President Allende's fight for the survival of his Marxist government against national and foreign capitalist interests than an indictment of his own policies.

19 The Report of the Royal Commission on Local Government in England, Cmnd. 4040 (HMSO, 1969), p. 125.

20 Interview: 'Towards machine intelligence,' *Science Journal* (September, 1968).

21 Carl J. Friedrich and Zbigniew K. Brzezinski, *Totalitarian Dictatorship and Autocracy* (New York, 1966).

22 *Science for People*, No 20 (BSSRS, February/March, 1973).

23 Frederick Engels and Karl Marx, *The German Ideology* (London, 1938), p. 7.

24 The idea that the concept of alienation is itself an ideological construct has most recently been argued by the French philosopher Louis Althusser.

25 Erich Fromm, *The Revolution of Hope: Towards a Humanised Technology* (New York, 1968), pp. 40, 41.

26 Melvin Seeman, 'On the meaning of alienation,' *American Sociological Review*, Vol. 24, (1959).

27 Robert Blauner, *Alienation and Freedom* (Chicago, 1964), p. 3.

28 Blauner, *ibid.*, p. 174.

29 For a graphic description of the conditions of the contemporary production line, see: Huw Beynon, *Working For Ford* (Harmondsworth, 1973).

30 W. A. Faunce, 'Automation in the Automobile Industry: some consequences for in-plant structure,' *American Sociological Review*, Vol 23, (1958).

31 B. Karsh, 'Work and Automation' in H. B. Jacobson and C. Roncek (eds.), *Automation and Society* (New York, 1959).

32 *The Guardian*, March 21, 1973 (my italics).

33 Mike Cooley, *Computer-Aided Design: Its Nature and Implications* (AUEW, London, 1972), p. 75.

34 For a discussion of this see, for example: Andrew Glynn and Bob Sutcliffe, *British Capitalism, Workers and the Profits Squeeze* (Harmondsworth, 1972).

35 Donald Schon, *Technology and Change: The New Heraclitus* (London, 1967), p. 194.

36 Jacques Ellul, *The Technological Society* (London, 1965), p. 14.

37 Herbert Marcuse, *One-Dimensional Man* (London, 1968), p. 131.

38 Theodore Roszak, *Making of a Counter Culture* (London, 1970), p. 9.

39 Jacob Bronowski, 'Technology and Culture in Evolution,' *Cambridge Review* (May 8, 1970).

40 Robin Clarke, 'Technology for an alternative society,' *New Scientist* (January 11, 1973).

41 Aldous Huxley, *Towards New Horizons*, by Pyarelal. Quoted by E. F. Schumacher. 'The Economics of Permanence,' *Resurgence*, Vol. 3, No. 1 (May/June, 1970).

42 Quoted in Schumacher, *op. cit.*

2 The Ideology of Industrialization

1 I am using the word 'ideology' in a manner broadly similar to the French sociologist, Henri Lefebvre, when he refers to it as 'a collection of errors, illusions and mystifications' which present 'an inverted, truncated, distorted reflection of reality'. See Henri Lefebvre, *The Sociology of Karl Marx* (London, 1966), p. 64.

2 T. S. Ashton, *The Industrial Revolution 1760–1830* (Oxford, 1948), p. 129.

3 Ashton, *ibid.*, p. 99.

4 André Gorz, 'Technical Intelligence and the Capitalist Division of Labour,' *Telos*, No. 12 (Summer, 1972).

5 Leslie A. White, quoted in Peter Hammond (ed.), *Cultural and Social Anthropology* (New York, 1964), p. 422.

6. Lynn White, Jr., *Mediaeval Technology and Social Change* (Oxford, 1962), p. 28.

7 White, *ibid.*, p. 76.

8 Frederick Engels, *The Condition of the Working Class in England* (London, 1969) (reprinted), p. 37.

9 Engels, *ibid.*, p. 55.

10 R. J. Forbes, *The Conquest of Nature: Technology and its Consequences* (Harmondsworth, 1971), p. 7.

11 Marshall McLuhan, *Understanding Media, The Extensions of Man* (London, 1967), p. 15.

12 McLuhan, *ibid.*, p. 197.

13 Joseph Needham, *Science and Civilisation in China*, Vols. I, II, III (Cambridge, various dates).

14 John F. Mee, 'Science and Management: Human Progress Twins,' *Advanced Management Journal* (October, 1964).

15 Lewis Mumford, *Technics and Civilisation* (London, 1934), p. 109.

16 Karl Marx and Frederick Engels, *The Manifesto of the Communist Party* (London, 1848).

17 Karl Marx, *Capital*, Volume 1 (London, 1930), p. 354.

18 Adam Smith, *An Inquiry into the Nature and Causes of the Wealth of Nations* (London, 1904), p. 5.

19 Smith, *ibid.*, p. 12.

20 Galbraith, *op. cit.*, p. 23.

21 Ashton, *op. cit.*, p. 57.

22 Ashton, *ibid.*, p. 17.

23 Marx and Engels, *op. cit.*

24 Marx, *op. cit.*, p. 394.

25 Marx, *ibid.*, p. 416.

26 V. I. Lenin, *The State and Revolution* (Peking, 1965), p. 58.

27 Quoted by Gustav Metzger, 'Automata in History,' *Studio International* (March, 1969).

28 Frederick Winslow Taylor, *Principles of Scientific Management* (New York, 1911), p. 59.

29 V. I. Lenin, *The Immediate Tasks of the Soviet Government*, in *Collected Works*, vol. 27 (Moscow, 1965), p. 259.

30 V. I. Lenin, 'The Taylor System – Man's Enslavement by the Machine,' in *Collected Works*, Vol. 20 (Moscow, 1964), p. 153.

31 *Science Growth and Society*, Organization for Economic Co-operation and Development (Paris, 1971), p. 7.

32 Alvin Toffler, *Future Shock* (London, 1970), p. 404.

33 W. T. Singleton, J. G. Fox, D. Whitfield (eds.), *Measurement of Man at Work* (London, 1971), p. 58.

34 Samir Amin, 'Development Strategies and Population Policies – Underpopulated Africa,' paper delivered to Unesco conference on *Population, Education, Development in Africa South of Sahara* (Dakar, 1971), p. 181.

35 Jürgen Habermas, *Towards a Rational Society* (London, 1971), p. 89. See also discussion on 'scientism' in Chapter 7.

36 John Jewkes, *Government and High Technology* (London, 1972).

37 Frederick Engels, *On Authority*, in *Die Neue Zeit*, Vol. XXXII, Band 1, 1913–1914, p. 39 (my italics).

38 Mikhail Drobyshev, 'Sociology of the tension between culture and science today.' Paper delivered to Unesco conference on *Culture and Science* (Paris, September 1971).

39 Resolution adopted by the Central Committee of the French Communist Party (PCF) at Argenteuil (March, 1966).

40 Claude Lévi-Strauss, *The Savage Mind* (London, 1966), p. 11.

41 E. E. Evans-Pritchard, *Witchcraft, Oracles and Magic Among the Azande* (London, 1936), p. 194.

42 Robin Horton, 'African Traditional Thought and Western Science,' in *Africa*, Vol. XXXVII (1967).

3 The Politics of Technical Change

1 David Landes, *The Unbound Prometheus: Technological Change and Industrial Development in Western Europe from 1750 to the Present* (Cambridge, 1969), p. 317.

2 See: William B. Kemp, 'The Flow of Energy in a Hunting Society,' *Scientific American*, Vol. 224, No. 3 (September, 1971).

3 Fredrick Barth, 'Nomadism in the Mountain and Plateau Regions of South West Asia,' in *Problems of the Arid Zone* (Paris (Unesco), 1960), p. 345.

4 Lévi-Strauss, *op. cit.*, p. 15.

5 Lévi-Strauss, *ibid.*, p. 17.

6 Horton, *op. cit.*

7 Joseph Ki-Zerbo, *La Civilisation Africaine d'Hier et de Demain* (Ouagadougou, Presses Africaines).

8 A. Touraine and others, *Workers' Attitude to Technical Change* (Paris (OECD), 1965), p. 8.

9 Mumford, *op. cit.*, p. 35.

10 A. N. Whitehead, *Science and the Modern World* (Cambridge, 1926), p. 19.

11 See: *Antiquarian Horology*, Vol. 1 (1954), p. 49.

12 White, *op. cit.*, p. 124.

13 See: A. C. Crombie, *Augustine to Galileo*, Volume 2 (London, 1961), p. 199. For a description of the political factors in feudal society leading to the replacement of hand-mills by water-mills, see: Stephen Marglin, *What do Bosses do?* (mimeographed, 1973).

14 See: John Read, *Prelude to Chemistry* (London, 1939), p. 30.

15 Marx, *op. cit.*, Vol. 1, p. 311.

16 Engels, *op. cit.*, p. 38. See also E. P. Thompson, 'Time, Work Discipline and Industrial Capitalism', *Past and Present*, No. 38 (December 1967).

17 Landes, *op. cit.*, p. 60.

18 See: Stephen Marglin, *op. cit.*

19 E. P. Thomson, *The Making of the English Working Class* (Harmondsworth, 1968), p. 308.

20 Neil McKendrick, 'Josiah Wedgwood and Factory Discipline,' *The Historical Journal*, Volume 4 (1961).

21 Landes, *op. cit.*, p. 60.

22 Ashton, *op. cit.*, p. 88.

23 R. M. Hartwell (ed.), *The Industrial Revolution* (Oxford, 1970), p. 31.

24 E. J. Hobsbawm, 'The Machine Breakers,' *Past and Present*, Volume 1 (February, 1952).

25 Samuel Butler, *Erewhon* (London, 1872), p. 180.

26 Landes, *op. cit.*, p. 115.

27 Thomson, *op. cit.* (1968), p. 222f.

28 Andrew Ure, *The Philosophy of Manufactures or an Exposition of the Scientific, Moral and Commercial Economy of the Factory System of Great Britain* (London, 1835), p. 365.

29 Ure, *ibid.*, p. 366.

30 Ure, *ibid.*, p. 368.

31 Quoted in Mumford, *op. cit.*

32 Marx, *op. cit.*, p. 464.

33 See: T. W. Schultz, 'Investment in Human Capital,' *American Economic Review*, 1962.

34 Landes, *op. cit.*, p. 117.

35 Marx, *op. cit.*, pp. 428–9.

36 Marx, *ibid.*, p. 453.

37 J. L. and Barbara Hammond, *The Town Labourer* (London, 1917), p. 306.

38 Hammond and Hammond, *ibid.*, p. 32.

39 Norman Ware, *The Industrial Worker, 1840–1860* (New York, 1924), p. xi.

40 Nathan Rosenberg, 'Technological Change,' in William Parker (ed.) *A New American Economic History* (New York, 1970).

41 Touraine et al., *op. cit.*, p. 38.

42 Touraine et al., *ibid.*, p. 11.

References and Notes

43 See: Robert F. Hoxie, *Scientific Management and Labour* (New York, 1915).

44 Alex Carey, 'The Hawthorne Studies: A Radical Criticism,' *American Sociological Review*, Volume 32 (1967).

45 Gorz, *op. cit.*

46 Andrew Glyn and Bob Sutcliffe, *British Capitalism, Workers and the Profits Squeeze* (Harmondsworth, 1972), p. 139.

47 Cooley, *op. cit.*, p. 76.

48 See: David McLellan, *Marx's Grundrisse* (London, 1971), p. 36.

49 Rosenberg, *op. cit.*,

50 Mumford, *op. cit.*, p. 58.

51 Quoted in Mumford, *ibid.*, p. 71.

52 Galbraith, *op. cit.*, p. 349.

53 Galbraith, *ibid.*, p. 350.

54 Herbert Marcuse, *One-Dimensional Man*, p. 135.

55 Lord Zuckerman, *Times Literary Supplement* (November 5, 1971).

56 Steven Rose, 'Can Science be Neutral?' *Proceedings of the Royal Institution*, Volume 45 (1972).

57 Mumford, *op. cit.*, p. 364.

58 Herbert Marcuse, *Eros and Civilisation* (London, 1969), p. 45.

4 Utopian Technology: some basic principles

1 Landes, *op. cit.*, pp. 32, 33.

2 See p. 67.

3 Karl Mannheim, *Ideology and Utopia* (London, 1936), p. 197.

4 John Todd, 'A Modest Proposal,' *The New Alchemy Institute Bulletin* (Spring, 1971).

5 Peter van Dresser, *A Landscape for Humans* (Albuquerque, New Mexico, 1972), p. 36.

6 Colin Moorcraft, 'Design for Survival,' *Architectural Design*, Volume 42 (July, 1972).

7 See p. 116.

8 Murray Bookchin, *Post-Scarcity Anarchism* (Berkeley, 1971), p. 132.

9 See, for numerous examples: Louis E. Davis and James C. Taylor (eds.) *The Design of Jobs* (Harmondsworth, 1972). Also Beynon, *op. cit.* For a critique of the way job enrichment can be used by management against the interests of

labour, see the pamphlet on job enrichment published by the British Society for Social Responsibility in Science.

10 See: Keith Paton, 'Work and Surplus,' *Anarchy*, No. 118 (December, 1970).

11 Paul and Percival Goodman, *Communitas* (New York, 1960), p. 155.

12 For a general discussion of self-management ideas, see: Ken Coates and Tony Topham, *Workers Control* (London, 1970); *Workers' Councils*, Solidarity Pamphlet No. 40.

13 Commoner, *op. cit.*, p. 33.

14 Claude M. Summers, 'The Conversion of Energy,' *Scientific American* (September, 1971).

15 See, for example: Peter Chapman, 'No overdrafts in the energy economy,' *New Scientist* (May, 17, 1973).

16 Chauncey Starr, 'Energy and Power,' *Scientific American* (September, 1971).

17 Summers, *op. cit.*

18 See: Bob Morgan, 'Sea Horse,' *Alternative Sources of Energy*, No. 10 (Spring, 1973).

19 For a description of the present state of methane gas technology, see: *Methane, Fuel of the Future*, published by Andrew Singer, Bottisham Park Mill, Cambridgeshire.

20 Bookchin, *op. cit.*, p. 129.

5 *Utopian Technology: commodity production and social organization*

1 F. M. C. Fourier, *Selections from the Work of Charles Fourier* (London, 1901), p. 93.

2 Bookchin, *op. cit.*, p. 118.

3 Lloyd Kahn, *Smart but not Wise: Further Thoughts on Domebook 2, Plastics and Whiteman Technology* (Bolinas, California, 1972).

4 Malcolm B. Wells, 'An ecologically-sound architecture is possible,' *Architectural Design*, Volume 42 (July, 1972), p. 434.

5 See: Alexander Pike, 'Cambridge Studies,' *Architectural Design*, Volume 42 (July, 1972), p. 441.

6 Quoted in *The Guardian* (April 10, 1973).

7 Umberto Nobile, 'Cyclists of the Air,' *New Scientist* (September 9, 1971).

8 See: Nicholas Pole, 'Watt's New in Motive Power,' *Your Environment*, Volume III, No. 3 (Autumn, 1972).

9 Andrew Jamison, *The Steam Powered Bicycle*, quoted by Pole, *op. cit.*

10 John Powles, paper given at conference on alternative technology, University College, London, February, 1972.

11 For a summary of many different approaches to the practice of medicine, see: *Health Manpower and the Medical Auxiliary* (London (ITDG), 1971).

12 Robin Clarke, 'Technology for an alternative society,' *New Scientist* (January 11, 1973).

13 Bookchin, *op. cit.*, pp. 112 and 220. The dangers in computer-based social planning have already been referred to in chapter 1 (see pages 28-29).

14 Paul Smoker, 'An Action Research Proposal for Global Networks,' *Peace and Conflict Research Programme Newsletter* (November, 1972).

15 'A Blueprint for Survival,' *The Ecologist*, Volume 2, No. 1 (January, 1972). The title was originally taken from the activities of the New Science Group which met briefly in London in the autumn of 1970 although with very different proposals in mind. (See: Michael Allaby, *The Eco-activists*, London, 1971, p. 6).

16 See: Hugh Thomas, 'Anarchist Agrarian Collectives in the Spanish Civil War,' in Martin Gilbert (ed.), *A Century of Conflict, 1850-1950* (New York, 1967). Also discussion in: Noam Chomsky, *American Power and the New Mandarins* (Harmondsworth, 1969).

17 See p. 106.

18 Peter van Dresser, *op. cit.*, p. 123.

19 Josiah Warren, 'Practical Details in Equitable Commerce,' quoted in M. S. Shatz (ed.) *The Essential Works of Anarchism* (New York, 1971), p. 442.

20 For a discussion of this relationship, see Chapter 7.

21 Jerome Ravetz, *Scientific Knowledge and Its Social Problems* (Oxford, 1971), pp. 424-431.

22 Jeremy Swift, 'Flaws in the Blueprint,' *New Scientist* (February 10, 1972).

23 Theodore Roszak, *Where the Wasteland Ends* (NewYork, 1973), p. 394.

6 Intermediate Technology and the Third World

1 See, in particular: W. W. Rostow, *The Stages of Economic Growth* (Cambridge, 1960).
2 Simon Kuznets, 'Underdeveloped countries and the Pre-Industrial Phase in the Advanced Countries,' in Otto Feinstein (ed.) *Two Worlds of Change* (New York, 1964), p. 8.
3 E. F. Schumacher, *Social and Economic Problems calling for the Development of Intermediate Technology*, mimeographed (undated). See also: E. F. Schumacher, *Small is Beautiful* (London, 1973).
4 A. Latham-Koenig, 'Intermediate Technology Development Group,' paper given to OECD conference (Paris, November 7–9, 1972).
5 Intermediate Technology Development Group Ltd., information sheet. undated.
6 ITDG Annual Report (1970/71).
7 Mrs Indira Gandhi, speech given at the inauguration of the National Committee on Environmental Planning and Co-ordination (Delhi, April 12, 1972).
8 Summary Proceedings, 1971 Annual Meeting of the World Bank, p. 21.
9 Schumacher. *op. cit.*
10 Hans Singer, paper given at conference on alternative technology, University College (London, February. 1973).
11 Raul Prebisch, statement to the third United Nations Conference on Trade and Development, Santiago, Chile on April 26, 1972.
12 C. Phipps, 'Technologists to lead in the Third World,' *New Scientist* (October 2, 1969).
13 Glenn E. Schweitzer, 'Towards a Methodology for Assessing the Impact of Technology in Developing Countries,' delivered to a seminar on Technology and Economics in International Development (Washington, May, 1972).
14 Hamza Alavi, 'The Post-Colonial State,' *New Left Review*, No. 74 (July/August, 1972).

15 Stephen Enke and Richard Brown, 'Economic Worth of Preventing Death at Different Ages in Developing Countries,' *Journal of Biosocial Science*, Vol. 4, No. 3 (July, 1972).

16 ITDG information leaflet (my italics).

17 See page 155f.

18 Hans Singer, *op. cit.*

19 Howard Pack, 'The Use of Labour Intensive Techniques in Kenyan Industry' (Washington, May, 1972) (see note 13).

20 Andre Gunder Frank, *Sociology of Development and Underdevelopment of Sociology* (London, 1971), p. 33.

21 Junta del Acuerdo de Cartagena, 'Transfer of Technology,' paper prepared for UNCTAD conference, Santiago, Chile (April 13, 1972).

22 Fernando Henrique Cardoso, 'Latin American Capitalism,' *New Left Review*, Number 74 (July/August, 1972).

23 Frank, *op. cit.*, p. 34.

24 L. T. Wells, Jr., 'Economic Man and Engineering Man: Choice of Technology in a Low Wage Country,' in *Economic Development Report* (Harvard Development Advisory Service, Autumn, 1972).

25 See: Genevieve Dean, 'China's Technological Development,' *New Scientist* (May 18, 1972). Also: Genevieve Dean and Manfredo Macioti, 'Scientific Institutions in China,' *Minerva*, Volume XI, No. 3 (July, 1973).

26 See: Hilary and Steven Rose, *BSSRS Newsheet* (March/April/May 1971).

27 Ivan Illich, *Celebration of Awareness* (London, 1971), p. 181.

28 William Bourke, 'Basic Vehicle for South-East Asia' (Washington, May, 1972). (See note 13).

29 Robert Macnamara, *op. cit.*, p. 20.

7 Myths and Responsibilities

1 Susanne K. Langer, *Philosophy in a New Key* (New York, 1951), p. 48.

2 See: David McLellan, *Marx's Grundrisse* (London, 1973), p. 156.

3 It is important, therefore, to distinguish between the processes of invention and innovation. 'Invention is the creation of a

new idea or technique. The innovation is the application of that new idea or technique to the actual process of production,' according to B. R. Williams, *Technology, Investment and Growth* (London, 1967), p. 33.

4 The basic concept of 'myth' used here is taken from *Mythologies* (London, 1972), by Roland Barthes, in particular his section on 'Myth Today'. 'What the world supplies to myth is an historical reality, defined, even if this goes back quite a while, by the way in which men have produced or used it; and what myth gives in return is a *natural* image of this reality,' writes Barthes (p. 141).

5 Jürgen Habermas, *Knowledge and Human Interests* (London, 1972), p. 4. The main difference between 'scientism' and the scientific (or technological) rationality of Herbert Marcuse is that whereas the latter is meant as an *explanation* of the way that social decisions are taken, the former represents the ideological interpretation or legitimation placed on these decisions, which are themselves interpreted as being taken according to a *political* rationality. For a discussion of the difference between the ideas of two philosophers, see 'Technology and Science as Ideology' in Jürgen Habermas, *Toward a Rational Society* (London, 1971).

6 For a critique of the conventional view, see: John Langrish, 'Does Industry Need Science?', *Science Journal* (December 1969).

7 For further discussion see: Gorz, *op. cit.*; Mike Hales, 'Management Science and the Second Industrial Revolution,' *Radical Science Journal*, No. 1 (1974); Robert Young, 'Darwinism and the Division of Labour,' *The Listener*, Vol. 88 (August 17, 1971).

8 Edmund Leach, 'Has Science a History?' *Journal of the Royal Society of Arts* (June, 1973).

9 Leibniz, *Monadology* (Oxford. 1898).

10 See: Jean Gimpel, *The Cult of Art* (London, 1969), Chapter 9.

11 Victor Ferkiss, *Technological Man, the Myth and the Reality* (London, 1969), p. 247.

12 Jacques Monod, *Chance and Necessity* (Harmondsworth, 1973). A somewhat more sophisticated discussion of the philosophical implications of contemporary research in biology

and genetics is given by Monod's co-Nobel Prize Winner, François Jacob in his *The Logic of Living Systems* (London, 1974). Jacob argues the need to introduce an awareness of social and political factors influencing the ideas of biologists into any discussion of the biological basis of evolution.

13 For further discussion of the ideological nature of the apparent confrontation between art and science, see: David Dickson, 'Beyond appearance: Some Critical Reflections on the Relationship Between Science and Art,' *Impact of Science on Society*, Vol. XXIV, No. 1 (January–March, 1974).

14 For a critique of technology assessment techniques, see: Brian Wynne, 'Superfix or superfixation,' *Science for People*, No. 24 (November, 1973).

15 *Science for People*, No. 20 (February/March, 1973).

16 Ann Rosenberg, 'Black Panthers in and on Science,' *New Scientist* (February 15, 1973).

17 Ernest Mandel, *The Changing Role of the Bourgeois University* (London, 1972).

18 See, for example, E. F. Schumacher, *Small is Beautiful* (London, 1973).

19 See: Michael F. D. Young (ed.), *Knowledge and Control* (London, 1971).

20 Marx, *op. cit.*, p. 382.

21 Robert Young, 'The Human Limits of Nature,' in Jonathan Benthall (ed.), *The Limits of Human Nature* (London, 1973), p. 263.

Index